Machining Composite Materials

Books are to be returned on or before
the last date below.

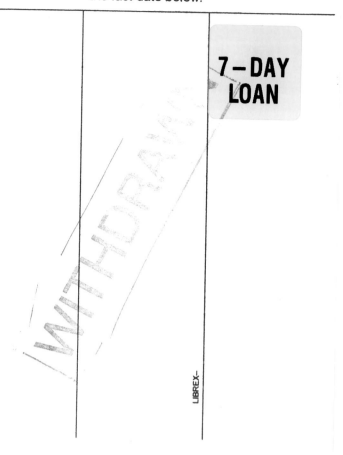

Machining Composite Materials

Edited by
J. Paulo Davim

First published in Great Britain and the United States in 2010 by ISTE Ltd and John Wiley & Sons, Inc.

Apart from any fair dealing for the purposes of research or private study, or criticism or review, as permitted under the Copyright, Designs and Patents Act 1988, this publication may only be reproduced, stored or transmitted, in any form or by any means, with the prior permission in writing of the publishers, or in the case of reprographic reproduction in accordance with the terms and licenses issued by the CLA. Enquiries concerning reproduction outside these terms should be sent to the publishers at the undermentioned address:

ISTE Ltd
27-37 St George's Road
London SW19 4EU
UK

www.iste.co.uk

John Wiley & Sons, Inc.
111 River Street
Hoboken, NJ 07030
USA

www.wiley.com

© ISTE Ltd, 2010

The rights of Paulo J. Davim to be identified as the author of this work have been asserted by him in accordance with the Copyright, Designs and Patents Act 1988.

Library of Congress Cataloging-in-Publication Data

Machining composite materials / edited by J. Paulo Davim.
    p. cm.
  Includes bibliographical references and index.
  ISBN 978-1-84821-170-4
  1. Composite materials. 2. Machining--Materials. 3. Manufacturing processes. I. Davim, J. Paulo.
  TA418.9.C6M25 2009
  620.1'18--dc22
                                                                                                    2009030939

British Library Cataloguing-in-Publication Data
A CIP record for this book is available from the British Library
ISBN: 978-1-84821-170-4

Printed and bound in Great Britain by CPI Antony Rowe, Chippenham and Eastbourne.

Table of Contents

Preface xi

Chapter 1. Mechanics and Modeling of Machining Polymer Matrix Composites Reinforced by Long Fibers 1
Liangchi ZHANG

  1.1. Introduction. 1
  1.2. Orthogonal cutting 2
    1.2.1. Surface roughness 4
    1.2.2. Cutting forces 7
    1.2.3. The bouncing back phenomenon 9
    1.2.4. Subsurface damage. 11
    1.2.5. Effect of curing conditions. 12
  1.3. Cutting force modeling. 13
    1.3.1. Region 1 14
    1.3.2. Region 2 17
    1.3.3. Region 3 19
    1.3.4. Total cutting forces 20
  1.4. Drilling. 21
    1.4.1. Observations. 23
    1.4.2. Effect of drilling conditions 28
      1.4.2.1. Drilling speed and feed speed 28
      1.4.2.2. Cutting speed and feed rate 30
      1.4.2.3. Ratio of cutting speed to feed speed 31
      1.4.2.4. Drilling force 34
      1.4.2.5. An empirical equation. 34

1.5. Abrasive machining. 35
1.6. Concluding notes 35
1.7. References 36

Chapter 2. Machinability Aspects of Polymer Matrix Composites 39
Franck GIROT, Luis Norberto LÓPEZ DE LACALLE, Aitzol LAMIKIZ, Daniel ILIESCU and Mª Esther GUTIÉRREZ

2.1. The machining of polymer composites. 40
2.2. Tools 41
  2.2.1. Tool materials 41
    2.2.1.1. Carbide tools 43
    2.2.1.2. Coatings for carbide tools 46
    2.2.1.3. Polycrystalline diamond (PCD). 47
    2.2.1.4. Diamond abrasive or diamond grit tools 49
  2.2.2. Tool types 50
2.3. Cutting mechanisms in composite materials 54
  2.3.1. Influence of fiber orientation on cutting 55
  2.3.2. Influence of fiber orientation on tool wear 60
  2.3.3. Influence of fiber orientation on cutting loads. 62
  2.3.4. Influence of fiber nature on cutting 64
2.4. Composite material damage due to machining 65
  2.4.1. Mechanical damage 66
    2.4.1.1. Fiber linting 66
    2.4.1.2. Fiber pullout and matrix decohesion. 66
    2.4.1.3. Delamination 67
  2.4.2. Thermal damage 68
  2.4.3. Chemical damage. 70
2.5. Milling of composite materials. 70
  2.5.1. GFRP routing 70
    2.5.1.1. Tool geometries. 70
    2.5.1.2. Routing conditions. 71
    2.5.1.3. Surface quality and tool lifetime 72
  2.5.2. Milling of CFRP 75
    2.5.2.1. Criterion for the wear measurement 78
    2.5.2.2. Cutting force and "specific cutting force" 80
    2.5.2.3. Test of different carbide substrates. 83
    2.5.2.4. Test of coatings. 86
    2.5.2.5. Tool geometry test 88
    2.5.2.6. Test of tool with diamond coating 88

         2.5.3. KFRP routing 89
            2.5.3.1. Tool geometries. 89
            2.5.3.2. Routing conditions 90
         2.5.4. Tool wear. 91
            2.5.4.1. Carbide and diamond-coated carbide tool
            wear behavior. 91
            2.5.4.2. PCD tool wear behavior 94
            2.5.4.3. Diamond abrasive tool wear behavior. 95
            2.5.4.4. Predictive tool wear model. 97
      2.6. Turning of composite materials 101
         2.6.1. Tool geometries. 101
         2.6.2. Turning conditions 103
         2.6.3. Surface quality 105
      2.7. Conclusions. 106
      2.8. Acknowledgments 107
      2.9. References 107

Chapter 3. Drilling Technology 113
Alexandre M. ABRÃO, Juan C. CAMPOS RUBIO, Paulo E. FARIA
and J. Paulo DAVIM

      3.1. Introduction. 113
      3.2. Standard and special tools. 117
      3.3. Cutting parameters 123
      3.4. Tool wear 125
      3.5. Drilling forces 131
      3.6. Surface integrity. 140
         3.6.1. Delamination 141
         3.6.2. Surface roughness 149
      3.7. Dimensional and geometric deviations. 153
      3.8. Conclusions. 157
      3.9. Acknowledgements. 159
      3.10. References. 159

Chapter 4. Abrasive Water Jet Machining of Composites . . 167
François CÉNAC, Francis COLLOMBET, Michel DÉLÉRIS
and Rédouane ZITOUNE.

      4.1. Introduction. 167
      4.2. Brief history of AWJT 168
      4.3. AWJ machining process 168

       4.4. AWJ cutting process   171
       4.5. Quality of the kerf.   172
       4.6. AWJ cutting of composite materials   173
       4.7. Applications   176
       4.8. Perspectives.   178
       4.9. AWJ milling of composite materials   178
       4.10. References.   180

Chapter 5. Machining Metal Matrix Composites   181
Alokesh PRAMANIK and Liangchi ZHANG

       5.1. Introduction.   181
       5.2. Conventional machining.   182
          5.2.1. Turning   182
          5.2.2. Drilling   184
          5.2.3. Grinding   187
          5.2.4. Milling   189
       5.3. Non-conventional machining.   190
          5.3.1. Electro-discharge machining   190
          5.3.2. Laser-beam machining   192
          5.3.3. Electro-chemical machining   193
          5.3.4. Abrasive water jet machining   194
       5.4. Tool–workpiece interaction.   195
          5.4.1. Evolution of stress field.   196
          5.4.2. Development of the plastic zone   199
          5.4.3. Comparison of experimental and FE simulation
          observations.   202
       5.5. Summary   203
       5.6. References   203

Chapter 6. Machining Ceramic Matrix Composites   213
Mark J. JACKSON and Tamara NOVAKOV

       6.1. Introduction.   213
       6.2. Electro-discharge machining of CMCs.   213
       6.3. Water jet machining of CMCs   226
       6.4. Laser machining of CMCs   227
       6.5. Ultrasonic machining of CMCs   234
       6.6. Application of CMCs: cutting tool inserts.   245

6.7. Review of various technologies for machining CMCs . . . 251
  6.8. References 253

List of Authors 257

Index 261

Preface

Recently, the application of composite materials has increased in various areas of science and technology due to their special properties, namely for their use in aircraft, automotive, defense and aerospace industries as well as other advanced industries. Machining composite materials is a rather complex task owing to their heterogenity, and the fact that reinforcements are extremely abrasive. In modern engineering, high demands are being placed on components made of composites in relation to their dimensional precision as well as to their surface quality. As a result of potential applications, there is a great need to understand the questions associated with the machining of composite materials.

This book aims to provide the fundamentals and the recent advances in the machining of composite materials (polymers, metals and ceramics) for modern manufacturing engineering.

Chapter 1 provides the mechanisms and modeling of machining polymer matrix composites reinforced by long fibers. Chapter 2 contains machinability aspects of polymer matrix composites. Chapter 3 covers drilling technology. Chapter 4 contains information on abrasive water jet machining of composites. Chapter 5 then focuses on the machining of metal matrix composites. Finally, Chapter 6 is dedicated to the machining of ceramic matrix composites.

This book can be used as a textbook for students in their final year of an undergraduate engineering course or those studying the subject of the machining of composite materials at the postgraduate level. This book can also serve as a useful reference for academics, manufacturing and materials researchers, manufacturing and mechanical engineers, as well as professionals in composite manufacturing and related industries. The scientific interest in this book is evident for many important centers of research, laboratories and universities in the world. Therefore, it is hoped that this book will encourage and enthuse others researching in this field of science and technology.

I would like to thank to ISTE-Wiley for this opportunity and for their professional support. Finally, I would like to thank all the authors of the chapters for their availability to work on this book.

J. Paulo Davim
University of Aveiro, Portugal
October 2009

Chapter 1

Mechanics and Modeling of Machining Polymer Matrix Composites Reinforced by Long Fibers

1.1. Introduction

Polymer matrix composites reinforced by long fibers (PMCRLF) are an important class of materials in advanced structural applications due to their lightweight, highly modulus and highly specific strength. However, because of their anisotropic and heterogenous nature, these materials are difficult to machine. Machined composite surfaces often contain damage such as delamination, cracks and fiber dislodgements [DAV 03, RAM 97, WAN 03].

To improve the surface integrity of machined surfaces, while maximizing the machinability of PMCRLF, significant investigations have been carried out to understand the mechanics of cutting, the effect of fiber orientation and PMCRLF fabrication conditions on the quality of machined surfaces using various machining processes such as orthogonal cutting [WAN 03], drilling [ZHA 01a] and grinding [HU 03, HU 04].

Chapter written by Liangchi ZHANG.

In terms of the study methodology, the investigations can generally be divided into three categories, experimental study focusing on the macro/microscopic behavior of PMCRLF [WAN 03], mechanics modeling [ZHA 01b] and numerical simulation that treats the PMCRLF as macroscopically anisotropic materials or focuses on the fiber-matrix interactions microscopically [MAH 01a, MAH 01b].

This chapter will discuss some advances in the investigations on the machining of PMCRLF in relation to orthogonal cutting and drilling.

1.2. Orthogonal cutting

Two commercial resin systems, the F593 and MTM56 prepregs, were used in the experiment. To investigate the effect of cutting conditions on machinability, the F593 prepregs were used to make unidirectional 4 mm-thick carbon/epoxy panels, cured under the pressure of 0.6 MPa at a temperature of 177°C for 2 hours. These panels, with the dimensions of 300 mm × 500 mm, were then cut into specimens with dimensions of 15 mm × 45 mm with desired fiber orientations for the cutting experiment.

The fiber orientation θ is defined clockwise with respect to the cutting direction, as shown in Figure 1.1. The cutting tools used were made from tungsten carbide with a clearance angle of 7° and rake angles from –20° to 40°. The cutting forces were measured by a three-dimensional (3D) dynamometer, Kistler 9257B. The cutting speed was fixed at 1m/min.

Table 1.1 lists the cutting conditions. To examine the effect of the ratio of depths of cut to the fiber diameter, the machinability at small depths of cut was also investigated.

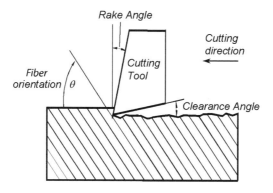

Figure 1.1. *A schematic of the orthogonal cutting of a PMCRLF specimen with unidirectional fibers orientated between 0° and 90°*

Fiber orientation (°)	0, 30, 60, 90, 120, 150
Rake angle (°)	−20, 0, 20, 40
Depth of cut (mm)	0.001, 0.050, 0.100

Table 1.1. *Cutting conditions for the F593 specimens*

Cure procedure	Temperature (°C)	Holding time (minute)
T1 (Under cure)	110	0
T2 (Standard cure)	120	10
T3 (Over cure)	120	20

Table 1.2. *Curing conditions for making the MTM56 specimens*

Fiber orientation (°)	0, 30, 60, 90, 120, 150
Rake angle (°)	0
Depth of cut (mm)	0.025, 0.050, 0.075, 0.100, 0.125, 0.150, 0.175, 0.200, 0.250
Cure procedure	T1, T2, T3

Table 1.3. *Cutting conditions for the MTM56 specimens*

In studying the influence of curing conditions, only the MTM56 prepregs were used because the F593 prepregs were not available on the market at that experiment stage. Table 1.2 lists the cure procedures under the pressure of 0.62 MPa. The temperature in each cure procedure was increased at 3°C/minute to the specified value, held at this value for the duration listed in Table 1.2 and then cooled down to room temperature.

Procedure T2 is a standard cure cycle recommended by the manufacturer of the MTM56 prepregs. Procedure T1 led to under-cured components and Procedure T3 gave rise to over-cured products. Table 1.3 lists the cutting conditions used for the MTM56 specimens.

The surface roughness of a machined surface was measured using a profilometer (Mitutoyo, Surftest 402 Series 178, cut-off = 0.8 mm). The morphology of the machined surface was observed using an optical microscope (Leica LEITZ DMRXE) and a scanning electron microscope (SEM) Philips XL-30.

1.2.1. *Surface roughness*

The surface roughness of the machined specimens made from F593 panels is presented in Figure 1.2, which shows the significant effect of the fiber orientation of the composite θ. It is clear that there exists a threshold, $\theta = 90°$, beyond which the surface roughness varies remarkably. At a given depth of cut smaller than the fiber diameter (7 μm – 9 μm), 1 μm for example, the surface roughness increases sharply when $\theta > 90°$ but decreases again when θ reaches 120°.

Before reaching the threshold of 90°, the change of the surface roughness is small, ranging from 0.6 μm to 1.2 μm, and the effects of rake angle and fiber orientation are relatively minor. At the most unfavorable fiber orientation of 120°, the surface roughness shows little dependence on the rake angle γ_0 with the best surface finish at $\gamma_0 = 20°$ and the worst at $\gamma_0 = -20°$.

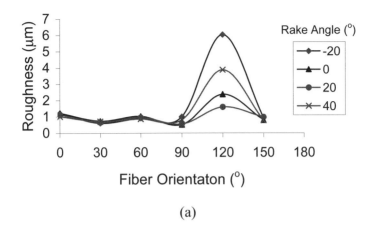

(a)

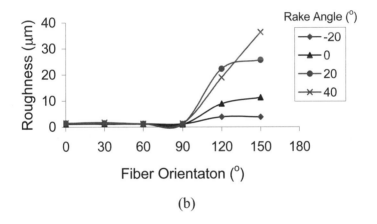

(b)

Figure 1.2. *Effect of fiber orientation on surface roughness (F593 specimens). The depths of cut were (a) 0.001 mm and (b) 0.050 mm*

6   Machining Composite Materials

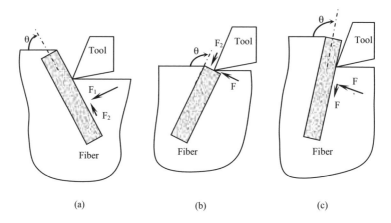

Figure 1.3. *Schematic cutting models*

When the depth of cut becomes larger than the fiber diameter, a different mechanism occurs. For instance, at the depth of cut of 50 μm the surface roughness does not decrease when θ is over 120°, as shown in Figure 1.2(b). The rake angle effect becomes greater and a sharper cutting tool (larger positive γ_0) produces a rougher surface. However, it is still true that θ = 90° is a critical angle, below which the effects of rake angle and fiber orientation are trivial. In this case, the surface roughness is in the range of 1 μm to 1.5 μm and is comparable to that when the depth of cut is smaller than the fiber diameter.

The above phenomenon may be explained by the variation of deformation mechanisms in the cutting zone when the depths of cut and fiber orientation change, as schematically illustrated in Figure 1.3 using a model with a single fiber. When θ is less than 90°, as shown in Figure 1.3(a), regardless of the depth of cut, the fiber is pushed by the tool (force F_1) in the direction perpendicular to the fiber axis and toward the workpiece subsurface. In this case, the fiber is better supported by the material behind and hence the bending of the fiber can be small. Meanwhile, the force component along the fiber axis (F_2) creates a tensile stress to make the fiber easier to break in the neighborhood of the cutting zone. This is because carbon fibers are brittle and can fracture easily under tension. As a result, the surface roughness and subsurface damage are small, as shown in Figure 1.2. When θ is greater than 90°, the situation becomes more complicated. When the depth of cut is less than $d\sin(\theta-90°)$, where d is the fiber diameter, i.e. when the tool is cutting at the end surface of the fiber as illustrated in Figure 1.3(b), the fiber is subjected to an axial compression. In this case, the fiber is unlikely to break when the surrounding epoxy, which is quite brittle, is fractured. Thus a machined surface normally has many protruded fibers, which results in a greater surface roughness. When the depth of cut becomes larger than $d\sin(\theta-90°)$, the tool exerts a different set of forces on the fiber, as shown in Figure 1.3(c). The pushing force (F_1) perpendicular to the fiber axis is towards the outside of the workpiece and hence the fiber receives less support from the surrounding materials, leading to more severe fiber bending and fiber-matrix debonding. This in turn causes a rougher surface finish and deeper subsurface damage. The mechanism of the surface roughness generation thus varies with the fiber orientation.

1.2.2. *Cutting forces*

For convenience, the cutting forces along and perpendicular to the cutting direction are called the horizontal and vertical forces, respectively. Figures 1.4 to 1.7 demonstrate the variation of the forces with the rake angle, fiber orientation and depth of cut.

The rake angle effect is not as significant if compared with the influence of the other two variables. At a small depth of cut (e.g., 1 µm), a rake angle between 0° and 20° gives rise to small cutting forces (Figure 1.4). At larger depths of cut (e.g., 50 µm), the horizontal force decreases slightly as the rake angle increases, except with the cases with θ = 120° and 150° (Figure 1.5). These are related to the resultant force variation when the fiber orientation and rake angle change. An interesting phenomenon is that the vertical forces at the fiber orientations of 120° and 150° decrease with the increment of rake angle (Figure 1.5(b)). This is understandable when the mechanics model in Figure 1.3(c) is recalled. At a greater fiber orientation, a larger rake angle and a higher depth of cut, the workpiece material applies a pulling force to the tool and hence the vertical cutting force becomes negative, as shown in Figures 1.5(b) and 1.7(b). When this occurs, the quality of the machined workpiece, including the surface roughness and subsurface damage, becomes poor, as shown in Figure 1.2.

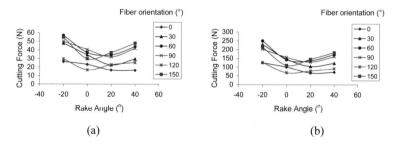

Figure 1.4. *Rake angle effect on (a) horizontal force, (b) vertical force (material: F593; depth of cut: 0.001 mm)*

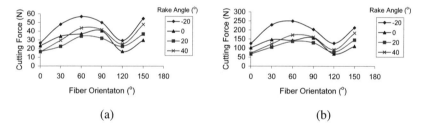

Figure 1.5. *Rake angle effect on (a) horizontal force, (b) vertical force (material: F593 ; depth of cut: 0.050 mm)*

The fiber orientation greatly influences the cutting forces as shown in Figures 1.6 to 1.7. At a small depth of cut (e.g., 1 μm, Figure 1.6), both the horizontal and vertical forces increase as θ increases, decrease when θ reaches 60°, and increase again after 120°. In the case with depths of cut of 50 μm and 100 μm, the horizontal force increases continuously until θ = 120° (Figure 1.7(a)).

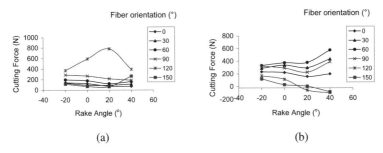

Figure 1.6. *Fiber orientation effect on (a) horizontal force, (b) vertical force (material: F593 ; depth of cut: 0.001 mm)*

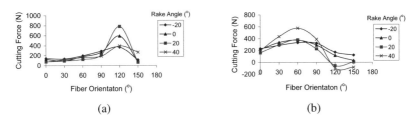

Figure 1.7. *Fiber orientation effect on (a) horizontal force, (b) vertical force (material: F593; depth of cut: 0.050 mm)*

1.2.3. *The bouncing back phenomenon*

It was found that the real depth of cut and nominal depth of cut are very different in cutting PMCRLF, as shown in Figure 1.8. This occurs when a part of the material in the cutting path was pushed down during the cutting but sprang back partially elastically after the tool passed by. Figure 1.9 shows the cutting force variation with the nominal depth of cut. Due to the bouncing back, the vertical cutting force increases with a distinguished slope corresponding to the increase of the bouncing back thickness (Figure 1.8). When the nominal depth of cut reaches a certain value, 100 μm in the current case, the magnitude of the bouncing back does not change much more and as a result the increasing rate of the normal cutting force also becomes very small. Figure 1.9 also shows that the dependence of the horizontal cutting force on the bouncing back is not strong, although the effect is still clear. All these indicate that the bouncing back is a key factor that contributes to the generation of the cutting forces.

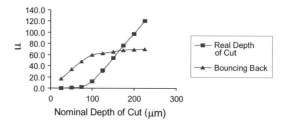

Figure 1.8. *Relationship between the nominal depth of cut, bouncing back and real depth of cut (material: MTM56; fiber orientation: 30°)*

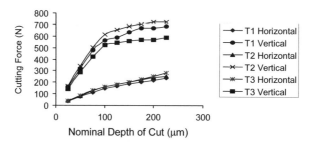

Figure 1.9. *Effect of the depth of cut and curing conditions on the cutting forces (material: MTM56; fiber orientation: 30°)*

It was found that the bouncing back magnitude is related to the radius of a cutting tool when all the other cutting conditions are the same. A series of measurements showed that when the fiber orientation θ is below 90°, the bouncing back magnitude is equal to or slightly greater than the tool radius. If θ is above 90°, the bouncing back magnitude can be up to more than twice the tool radius, depending on the θ value. As discussed in the previous section and illustrated in Figure 1.3, this is because at a greater θ value, many fibers were pushed to bend (but not break at the cutting point). When the cutting tool passed by, these fibers recovered elastically so that the bouncing back magnitude becomes large.

1.2.4. Subsurface damage

Under some specific cutting conditions, the subsurface of a machined specimen can be damage-free, but under other conditions, debonding and fiber-breakage can take place easily, as shown in Figure 1.10. The subsurface damage is related to the depth of cut, fiber orientation and rake angle. The observation shows that in general a smaller depth of cut generates less subsurface damage. At a larger depth of cut (e.g., 50 μm and 100 μm), the subsurface damage becomes more sever when the fiber orientation is between 120° and 150°. This explains why the surface roughness under these cutting conditions is also high. The fiber orientation plays a key role here. For instance, when θ = 150°, subsurface cracks appear only at the rake angle of 40°; but when θ = 120°, cracking always occurs regardless of the rake angle value. It is interesting to note that under different depths of cut, e.g. 50 μm and 100 μm, the subsurface has a similar feature if the fiber orientation and rake angle are the same.

Figure 1.10. *Microstructure in the subsurface of F593 specimens (fiber orientation: 120°, depth of cut: 0.100 mm). Top: rake angle = –20°; bottom: rake angle = 20°*

The SEM examination of the machined surfaces leads to the conclusions that are in agreement with the surface roughness observations. When θ ≤ 90°, a machined surface is not influenced much by the fiber orientation, depth of cut or rake angle.

1.2.5. Effect of curing conditions

The mechanical properties of the F593 panels were provided by the manufacturer, as listed in Table 1.4. The tensile tests on MTM56 panels were conducted in the authors' laboratory and the results are listed in Table 1.5. Clearly, the mechanical properties of the MTM56 panels are affected by the curing conditions. The best tensile strength was obtained by the standard curing procedure. The under cure leads to 90% of the tensile strength from the standard cure in the fiber direction and 75% in the transverse direction. The over cured specimen has only 76% of the tensile strength in the fiber direction and 60% in the transverse direction. On the other hand, however, the best elastic modulus was obtained by the under cure, which, compared to that from the standard cure, is 105% in the fiber direction and 111% in the transverse direction. The over cure procedure gives 91% and 98%, respectively. Curing conditions do not show a noticeable effect on the surface roughness after machining.

Tensile strength (MPa)	1331
Tensile modulus (GPa)	120.0
Compression strength (MPa)	1655
Compression modulus (GPa)	114.5

Table 1.4. *Mechanical properties of the F593 specimens*

Cure procedure	Type	Tensile strength (MPa)		Tensile Modulus (GPa)	
T1	Longitudinal	1623	90%	275.9	105%
T1	Transverse	21.3	75%	25.7	111%
T2	Longitudinal	1806	100%	261.6	100%
T2	Transverse	28.5	100%	23.2	100%
T3	Longitudinal	1385	76%	236.9	91%
T3	Transverse	17.0	60%	22.8	98%

Table 1.5. *Mechanical properties of the MTM56 panels with curing conditions*

The effects of the degree of cure on the cutting forces are shown in Figure 1.9. It can be seen that the horizontal cutting force is generally unaffected by cure conditions. The variation of the vertical cutting force with the cure conditions is similar to that of the tensile strengths of the materials. The degree of cure alters the microstructure of the MTM56 panels. Compared with the standard cure, both the under cure and over cure introduce subsurface defects during the manufacturing of composites. However, no other significant differences were found. This demonstrates the reason that the cutting forces are weak functions of the curing conditions.

1.3. Cutting force modeling

Based on the above understanding, the mechanics modeling of cutting needs to be conducted differently when $\theta \leq 90°$ or when $\theta > 90°$. This section will discuss the case of $\theta \leq 90°$ [ZHA 01b]. The reader may try to obtain a solution for the case with $\theta > 90°$.

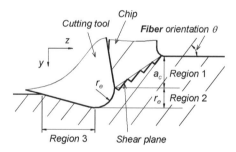

Figure 1.11. *Definitions of the cutting variables and deformation zones when the fiber orientation is smaller than 90°*

The deformation mechanisms demonstrated in the last section suggest that the cutting zone in mechanics modeling should be divided into three distinct regions as illustrated in Figure 1.11. Region 1 has a depth of a_c, bounded by the starting point of the tool nose according to the experiment. Region 2 covers the whole domain under the tool nose, having a depth equal to the nose radius, r_e. Region 3 starts from the lowest point of the tool. We assume that the total cutting force can be calculated by adding up the forces in all three regions, i.e., we

14   Machining Composite Materials

assume that the principle of superposition applies. For convenience, the positive directions of the forces are taken to be in the positive *y*- and *z*-directions defined in Figure 1.11.

1.3.1. Region 1

Chipping in this region is similar to a normal orthogonal cutting with a sharp cutting tool. Figure 1.12(a) shows a cutting force diagram, where AB is a theoretical shear plane, formed by many micro events of the cross-section fracture of fibers, along AC, and fiber-matrix debonding, along CB. In this region, the resultant force R is equal and opposite to the resultant force R', which consists of a shear force and a normal force acting on the shear plane AB. The shear force can be further resolved into Fs_1, which cuts the fibers along CA, and Fs_2, which cuts the matrix or delaminates the fiber-matrix interface along BC, parallel to the fiber axis.

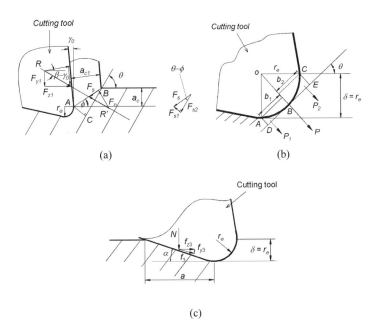

Figure 1.12. *(a) The cutting force diagram in Region 1. (b) The contact between the tool nose and the workpiece material in Region 2. (c) The contact in Region 3*

Fs_1 and Fs_2 can be written as

$$\begin{cases} F_{s1} = F_s \cdot sin(\theta - \varphi) \\ F_{s2} = F_s \cdot cos(\theta - \varphi) \end{cases} \quad [1.1]$$

where θ is the fiber orientation varying from 0° to 90° and ϕ is the shear plane angle to be determined. On the other hand, if τ_1 and τ_2 are the shear strengths of the work material in *AC* and *BC* directions, respectively, the two shear forces can be expressed as

$$\begin{cases} F_{s1} = \tau_1 \cdot l_{AC} \cdot h \\ F_{s2} = \tau_2 \cdot l_{BC} \cdot h \end{cases} \quad [1.2]$$

where l_{AC} is the total length of all the cross-section fractures along *AC*, l_{BC} is the total length of all the matrix fractures along *BC* and h is the thickness of the workpiece material perpendicular to the *yz*-plane. Therefore, by equating equations [1.1] and [1.2], we obtain

$$\frac{l_{BC}}{l_{AC}} = \frac{\tau_1}{\tau_2 \cdot \tan(\theta - \phi)} \quad [1.3]$$

According to Figure 1.12(a), we also have

$$l_{BC} \cdot sin\theta - l_{AC} \cdot cos\theta = a_c, \quad [1.4]$$

where a_c is the real depth of cut. Equations [1.3] and [1.4] then give rise to

$$l_{AC} = \frac{a_c}{\dfrac{\tau_1 \cdot \sin\theta}{\tau_2 \cdot \tan(\theta - \phi)} - \cos\theta} \quad [1.5]$$

Therefore, using equations [1.1], [1.2] and [1.5], the total shear force F_s in Region 1 can be calculated by

$$F_s = \frac{\tau_1 \cdot h \cdot a_c}{\frac{\tau_1}{\tau_2} \cdot \cos(\theta - \phi) \cdot \sin\theta - \sin(\theta - \phi) \cdot \cos\theta} \quad [1.6]$$

Since

$$F_n = F_s \cdot \tan(\phi + \beta - \gamma_o), \quad [1.7]$$

where β is the friction angle on the rake face, we have

$$\begin{pmatrix} F_{z1} \\ F_{y1} \end{pmatrix} = \begin{pmatrix} \sin\phi & \cos\phi \\ \cos\phi & -\sin\phi \end{pmatrix} \begin{pmatrix} F_n \\ F_s \end{pmatrix} \quad [1.8]$$

where F_{y1} and F_{z1} are vertical and horizontal cutting forces. Hence the substitution of equations [1.6] and [1.7] into equation [1.8] yields the cutting forces in Region 1:

$$\begin{cases} F_{z1} = \tau_1 \cdot h \cdot a_c \cdot \dfrac{\sin\phi \cdot \tan(\phi + \beta - \gamma_o) + \cos\phi}{\dfrac{\tau_1}{\tau_2} \cdot \cos(\theta - \phi) \cdot \sin\theta - \sin(\theta - \phi) \cdot \cos\theta} \\ \\ F_{y1} = \tau_1 \cdot h \cdot a_c \cdot \dfrac{\cos\phi \cdot \tan(\phi + \beta - \gamma_o) - \sin\phi}{\dfrac{\tau_1}{\tau_2} \cdot \cos(\theta - \phi) \cdot \sin\theta - \sin(\theta - \phi) \cdot \cos\theta} \end{cases} \quad [1.9]$$

To calculate the forces using equation [1.9], ϕ needs to be determined. According to the general cutting mechanics, we have

$$\tan\phi = \frac{r_c \cdot \cos\gamma_o}{1 - r_c \cdot \sin\gamma_o},$$

where γ_o is the rake angle of the tool and

$$r_c = \frac{a_c}{a_{c1}},$$

in which a_{c1} is the chip thickness. As an FRP in cutting behaves like a typical brittle material, it is reasonable to let $r_c = 1$. Therefore,

$$\phi \approx \tan^{-1}\left(\frac{\cos \gamma_o}{1-\sin \gamma_o}\right)$$

1.3.2. Region 2

The deformation in Region 2 is caused by the tool nose, which can be viewed as the deformation under a cylindrical indenter, as shown in Figure 1.12(b). Since *DB* and *BE* are generally unequal when the fiber orientation θ varies, the tool nose indentations by surfaces *AB* and *BC* need to be considered separately. Again, we assume that the principle of superposition applies. Using the indentation mechanics of a circular cylinder in contact with a half-space, the indentation force on the tool nose can be calculated approximately by adding up half of the indentation forces on arc lengths 2*AB* and 2*BC*. We therefore obtain

$$\begin{cases} P_1 = \frac{1}{2} \cdot \frac{b_1^2 \cdot \pi \cdot E^* \cdot h}{4 \cdot r_e}, \\ P_2 = \frac{1}{2} \cdot \frac{b_2^2 \cdot \pi \cdot E^* \cdot h}{4 \cdot r_e}, \end{cases} \qquad [1.10]$$

where P_1 and P_2 are the indentation forces, perpendicular to the fiber axis, that the tool nose exerts on *AB* and *BC* respectively. E^* is the effective elastic modulus of the workpiece material in the direction of P_1 and P_2, h is the thickness of the workpiece, r_e is the nose radius and

b_1 and b_2 are the widths of the contact arcs AB and BC, which can be calculated as

$$\begin{cases} b_1 = r_e \cdot \sin\theta \\ b_2 = r_e \cdot \cos\theta \end{cases} \qquad [1.11]$$

The effective elastic modulus E^* is defined by

$$E^* = \frac{E}{1-v^2} \qquad [1.12]$$

where E is the Young's modulus of the workpiece material in the direction of OP in Figure 1.12(b), and v is the minor Poisson's ratio. The resultant force P is therefore

$$P = P_1 + P_2. \qquad [1.13]$$

It must be pointed out that equation [1.10] is based on the contact mechanics of elastic deformation. In a cutting, however, the deformation of the workpiece material under the pressing of the tool nose must have introduced micro-cracking and matrix failure. To take such micro-effects into account, the resultant force P can be approximately modified by

$$P_{real} = K \cdot P, \qquad [1.14]$$

where P_{real} is called the real resultant force in Region 2 in which the coefficient K is a function of fiber orientation, i.e. $K = f(\theta)$, to be determined by experiment. Combining equations [1.10] with [1.14], we obtain

$$\begin{cases} F_{y2p} = P_{real} \cdot \cos\theta \\ F_{z2p} = P_{real} \cdot \sin\theta \end{cases}$$

When the friction coefficient is μ, the frictional force

$$f_{real} = P_{real} \cdot \mu,$$

can be resolved as

$$\begin{cases} f_{y2} = P_{real} \cdot \mu \cdot \sin\theta \\ f_{z2} = P_{real} \cdot \mu \cdot \cos\theta \end{cases}$$

Finally, the total cutting forces in Region 2 become

$$\begin{cases} F_{y2} = P_{real} \cdot (\cos\theta - \mu \cdot \sin\theta) \\ F_{z2} = P_{real} \cdot (\sin\theta + \mu \cdot \cos\theta) \end{cases} \quad [1.15]$$

1.3.3. *Region 3*

In this region, the contact force between the clearance face and the workpiece material is caused by the bouncing back of the workpiece material. For simplicity, assume that the bouncing back is complete, i.e. the height of the bouncing back is equal to the thickness of Region 2, r_e. Thus the contact length a in Region 3, as illustrated in Figure 1.12(c), can be obtained as

$$a = \frac{r_e}{\tan\alpha} \quad [1.16]$$

where α is the clearance angle of the tool. Using the contact mechanics between a wedge and a half-space, the total force N can be calculated by

$$N = \frac{1}{2} \cdot a \cdot E_3 \cdot \tan\alpha \cdot h \quad [1.17]$$

where E_3 is the effective modulus of the workpiece material in Region 3, which must be smaller than that of the original workpiece material because the material in this region has been damaged during the deformation experienced in Region 2 and thus has become weaker. Using equation [1.16], we obtain

$$N = \frac{1}{2} \cdot r_e \cdot E_3 \cdot h \cdot$$

The friction force f_3 between the clearance face of the cutting tool and the workpiece material is μN and can also be resolved into y- and z-directions to obtain f_{y3} and f_{z3}. Hence, the cutting forces in Region 3 are

$$\begin{cases} F_{y3} = \frac{1}{2} \cdot r_e \cdot E_3 \cdot h \ (1 \quad \mu \cdot \cos\alpha \cdot \sin\alpha) \\ F_{z3} = \frac{1}{2} \cdot r_e \cdot E_3 \cdot h \cdot \cos^2\alpha \end{cases}$$

1.3.4. *Total cutting forces*

The total forces, F_z and F_y, are the summation of the corresponding components from the above three regions, i.e.

$$\begin{cases} F_y = F_{y1} + F_{y2} + F_{y3} \\ F_z = F_{z1} + F_{z2} + F_{z3} \end{cases} \qquad [1.18]$$

where F_{yi} and F_{zi} (i = 1, 2, 3) are defined in equations [1.9], [1.15] and [1.18], respectively. In this mechanics model, the parameters to be determined by experiment, when a workpiece material is given, are τ_1, τ_2, β, E, v, μ, E_3 and K.

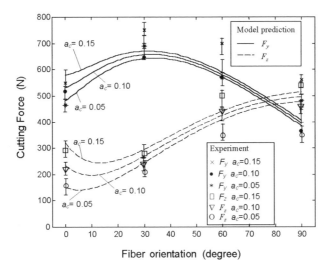

Figure 1.13. *Comparison between model prediction and experimental measurements when the depth of cut and fiber orientation change (material: MTM56, $E_3 = 5.5 GPa$). With these materials it was found that $\tau_1 = 90 mpa$, $\tau_2 = 20 mpa$, $\beta = 30°$, $\mu = 0.15$, $E = 10 GPa$, $v = 0.026$ and $h = 4m$*

Figure 1.13 shows the comparison between the model prediction and experimental measurements. It can be seen that although the model involves simplifications and assumptions as presented in the previous section, it predicts the cutting force variation nicely when the cutting parameters, such as the depth of cut, fiber orientation and rake angle, change. This means that the model has captured the major deformation mechanisms in cutting the composites.

1.4. Drilling

Similar to orthogonal cutting, drilling can cause significant damage to polymer matrix composites reinforced by long fibers. Exit defect is one of the major causes of damage. The degree of the defect represents the drilling quality, characterized by the so-called spalling at a hole exit. When drilling a PMCRLF plate with a twist drill made of high-speed steel, it has been found that the spalling damage increases with the feed rate but decreases with the spindle speed.

However, the effect of the feed rate is often greater than that of the spindle speed. Some studies reported that the degree of spalling may be minimized by arranging an aluminum or Bakelite plate under the PMCRLF panel subjected to drilling. In addition, to reduce the exit spalling, it is necessary to control the thrust force, especially at the stage when the chisel edge is penetrating the exit surface of a PMCRLF plate. This section will discuss a way of modeling such exit defects [ZHA 01a].

The drilling experiment for the exit defect was conducted using six spindle speeds and five feed speeds, as listed in Table 1.6. The drills used were 4-facet point solid carbide drills of diameters $\phi 4.8$ mm, $\phi 5.5$ mm, $\phi 6$ mm, respectively. The workpieces were made of two types of PMCRLF plates, unidirectional carbon fiber-reinforced plastics and multi-directional carbon fiber-reinforced plastics, as detailed in Table 1.7.

Spindle speed (rpm)	3,000, 6,000, 9,000, 12,000, 18,000, 24,000
Feed speed (mm/min)	24, 44, 66, 91.2, 120.8
Drill material and geometry	Material: YG6X carbide Geometry: 4-facet point Diameter: $\phi 4.8$ mm, $\phi 5.5$ mm, $\phi 6$ mm

Table 1.6. *Drilling parameters*

Multi-directional PMCRLF	
Carbon fiber	T300 of 7 μm in diameter
Matrix	Epoxy resin
Fiber direction	[+45°/0°/-45°/+45°/0°/90°/-45°/0°/90°/0°]s
Fiber volume content	60%
Workpiece thickness	2.5 mm
Unidirectional PMCRLF	
Carbon fiber	T300, diameter 7 μm
Matrix	Epoxy resin
Fiber direction	0°
Fiber volume content	60%
Workpiece thickness	2 mm

Table 1.7. *Workpiece materials*

As illustrated in Figure 1.14, parameter l, the average of the spalling lengths at the two sides of the hole exit, l_1 and l_2, i.e.,

$$l = \frac{l_1 + l_2}{2} \qquad [1.19]$$

is defined to assess the magnitude of spalling.

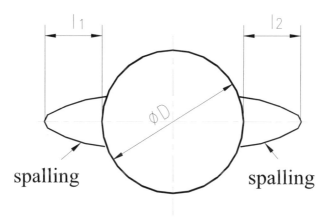

Figure 1.14. *The spalling*

1.4.1. *Observations*

It was found that the exit defect generally consists of two types of damage, spalling and fuzzing. The following results were obtained by examining the drilling process and the drilled workpieces:

1) a spalling damage develops along the fiber direction and its dimension is normally bigger than that of the accompanying damage by fuzzing;

2) the spalling is developed in two phases, the chisel edge action phase and the cutting edge action phase. The first phase begins when the thrust force of the chisel edge onto the exit surface reaches a critical value and ends when the chisel edge just penetrates the plate.

24 Machining Composite Materials

By examining the photographs of the exit surfaces and the finished workpieces, it was found that the chisel edge has a strong effect on the formation of the spalling. A small bulge emerges first in the vicinity of the drilling axis and then develops along the fiber direction of the exit surface (Figure 1.15.(a)). When the bulge grows to a certain degree, the surface layer splits open, the chisel edge penetrates and the second phase, cutting edge action phase, starts. The spalling damage initiated in the first phase develops further due to the continuous pushing (Figure 1.15.(b)) and twisting (Figure 1.15(c)) of the cutting edge.

The chisel edge cuts the workpiece material with a big negative rake angle and generates over 50% of the thrust force. Thus the chisel edge plays a key role. The experiment shows that at the moment of the chisel edge penetration, the spalling has already grown to a large part of its final size. Figure 1.16(a) shows a spalling at the chisel edge penetration and Figure 1.16(b) shows the spalling after drilling. We can easily confirm the above observations.

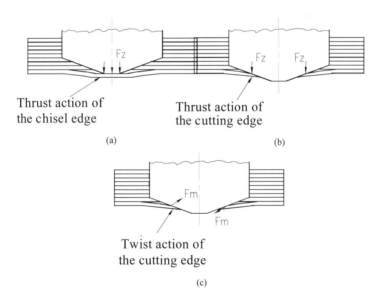

Figure 1.15. *Schematic of the formation process of the spalling defect*

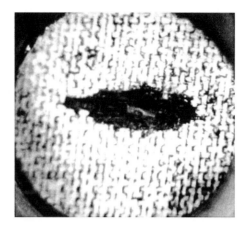

(a) spalling at the chisel edge penetration

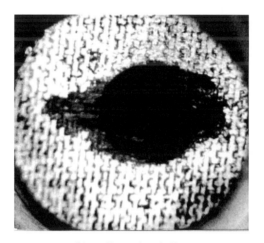

(b) spalling after drilling

Figure 1.16. *Spalling at the exit surface generated by a ϕ6 carbide drill with a drilling speed of 3,000 rpm and a feed speed of 91.2 mm/min. The workpiece was a multi-directional PMCRLF plate with the fiber orientation of 0° in its surface layer*

26   Machining Composite Materials

Figure 1.17 qualitatively describes the growth of the spalling damage in the two phases. It shows that the growing rate in the first phase is greater than the growing rate in the second phase.

If the drilling parameters, such as the feed rate, are appropriately chosen, the drilling force, particularly the thrust force, will become very small at the end of the first phase and the spalling dimension in the phase will be less than the hole diameter (see the dotted curve in Figure 1.17).

Then in the second phase, the cutting edges will remove the spalling formed in the first phase and after drilling the hole exit will become spalling-free.

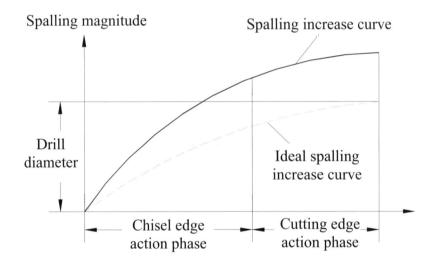

Figure 1.17. *Development of a spalling defect in two phases*

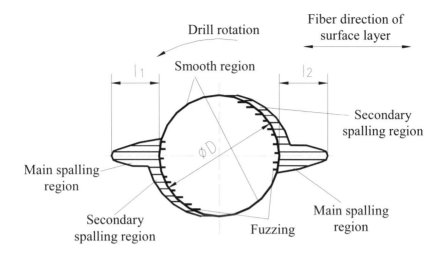

Figure 1.18. *Characteristics of an exit defect*

It is found that an exit defect can be characterized by a main region and a secondary region, as illustrated in Figure 1.18. The main region, which has the maximum spalling, appears radically in the fiber direction as shown. The secondary region is small and is generated in the cutting phase, in which the included angle between the fiber direction and that of the cutting speed is acute.

In addition, there often exist a small amount of fibers that are not cut off neatly at the edge of the hole in both the main and secondary spalling regions. This defect is called fuzzing, formed by the following causes:

− the fiber is not easy to be cut off in the region where the included angle between the fiber direction and that of the cutting speed is acute;

– the exterior of the surface layer at the hole exit is a free surface so that the fibers are not subjected to shear deformation.

Spalling and fuzzing usually co-exist and both their magnitudes have the same variation tendency, i.e. the bigger the spalling the more severe the fuzzing and vice versa. However, when the spalling decreases to a certain extent, the fuzzing disappears.

1.4.2. *Effect of drilling conditions*

1.4.2.1. *Drilling speed and feed speed*

Figure 1.19 shows the size variation of the spalling defect with the drilling speed n and feed speed v_f for the PMCRLF plates with unidirectional and multi-directional fiber-reinforcements respectively. It can be seen that the defect size decreases with n but increases with v_f, no matter what kind of PMCRLF is. This is similar to the drilling force variation.

For the multi-directional PMCRLF, at a high feed speed of $v_f = 66{\sim}120.8$ mm/min, the spalling size decreases quickly, especially at the lower drilling speed. This effect becomes negligible when the feed speed v_f becomes lower, e.g. between 24 and 44 mm/min. The same behavior exists in drilling the unidirectional PMCRLF.

It is thus clear that there are two ways to reduce spalling, i.e. using a smaller feed speed or operating at a high drilling speed. However, a lower feed speed leads to a lower machining efficiency. Thus, an operation with a high drilling speed is preferable. By comparing Figure 1.19(a) with 1.19(b), it can also be seen that the spalling in the unidirectional PMCRLF is generally bigger than that in the multi-directional PMCRLF under identical cutting conditions. This is understandable because in the former composite the constraints among fibers are much weaker.

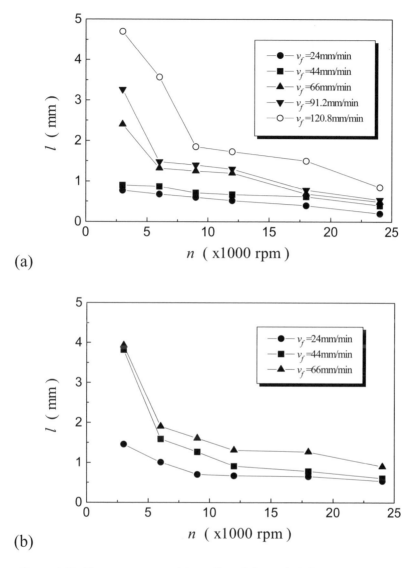

Figure 1.19. *The size variation of the spalling defect with drilling speed and feed speed. The drill diameter is $\phi 5.5$ mm. (a) Multi-directional PMCRLF; (b) unidirectional PMCRLF*

1.4.2.2. *Cutting speed and feed rate*

Figure 1.20 shows the spalling variation with cutting speed v and feed rate f when drilling multi-directional PMCRLF and unidirectional PMCRLF respectively, where v is the linear speed at the outer edges of the cutting lip.

It can be seen that the size of the spalling defect increases with the feed rate. It is also interesting to note that all the experimental data with various cutting speeds (51.8~432m/min) collapse to a line and this is particularly true when f is low. This seems to conclude that the feed rate is a dominant variable for spalling defects.

A similar phenomenon was observed in the variation of the drilling force. This will become clearer if our discussion in the previous section is recalled, i.e. spalling is caused by the chisel and cutting edge actions and drilling forces (thrust force and torque) reflecting the overall response of the workpiece material in the two phases.

By ignoring the influence of cutting speed, the linear relationship between the feed rate and the spalling can be expressed as two empirical equations when a linear regression is carried out, i.e.

$$l = 0.26 + 0.11 \cdot f \text{ for multi-directional PMCRLF;}$$

$$l = 0.34 + 0.18 \cdot f \text{ for unidirectional PMCRLF}$$

where l is the spalling size in millimeter defined by equation [1.19] and f is the feed rate in μm/rpm.

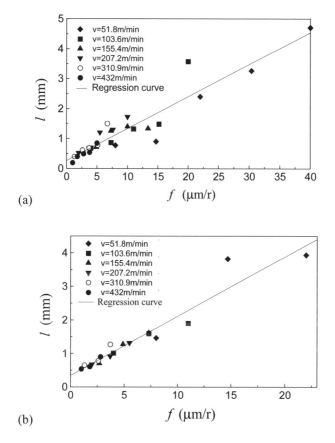

Figure 1.20. *The effects of the cutting speed and feed rate on the size of the spalling defect. The drill diameter was φ5.5 mm. (a) Multi-directional PMCRLF; (b) unidirectional PMCRLF*

1.4.2.3. *Ratio of cutting speed to feed speed*

Figures 1.21 and 1.22 show the spalling variation with the ratio of cutting speed to feed speed, v/v_f, when drilling multi-directional PMCRLF and unidirectional PMCRLF respectively, where Figures 1.21(a) and 1.22(a) have linear plotting axes and Figures 1.21(b) and 1.22(b) have log-log axes. From Figures 1.21(a) and 1.22(a) it can be seen that under different feed speeds the experimental data are all distributed in the vicinity of a curve. With the increase of v/v_f, the spalling generally decreases regardless of the type of PMCRLF, i.e. unidirectional or multidirectional. When v/v_f is smaller than 4,000, the spalling size l decreases more quickly. Further increasing the ratio does not make a noticeable change to l. A similar phenomenon also exists in the variation of the drilling force. A critical value of v/v_f is therefore approximately 4,000, which is helpful for selecting drilling parameters in production. When v/v_f is in the regime beyond the critical value, the spalling size generated will be small.

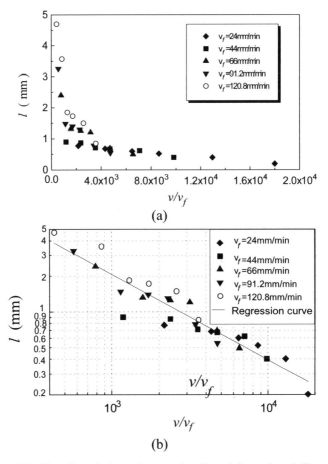

Figure 1.21. *The effect of v/v_f on the size of spalling defects when drilling multi-directional CFRP plates using a drill of $\phi 5.5$ mm. (a) Linear diagram; (b) log-log diagram*

Using the log-log diagrams, Figures 1.21(b) and 1.22(b), the empirical relationship between v/v_f and l can also be expressed as

$$l = 302 \cdot (v/v_f)^{-0.72} \text{ for multi-directional PMCRLF;}$$

$$l = 288.4 \cdot (v/v_f)^{-0.66} \text{ for unidirectional PMCRLF,}$$

where l is in mm.

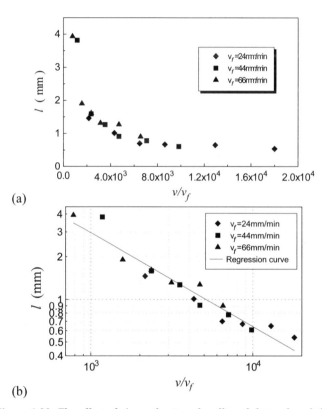

Figure 1.22. *The effect of v/v_f on the size of spalling defects when drilling unidirectional PMCRLF plates using a drill of $\phi 5.5$ mm. (a) Linear diagram; (b) log-log diagram*

1.4.2.4. *Drilling force*

Figure 1.23 shows the effect of the thrust force when drilling multi-directional PMCRLF. With the increase of the thrust force, the spalling size increases in quadratic manner as

$$l = 0.76 - 0.04 F_z + 0.001 F_z^2$$

where *l* is in mm and F_z is the thrust force in N.

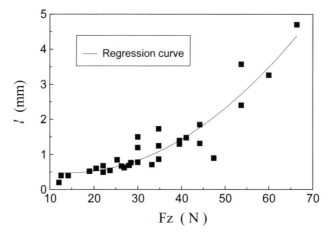

Figure 1.23. *Relationship between the spalling size and the thrust force. (ϕ5.5 mm solid carbide drill, multi-directional PMCRLF)*

1.4.2.5. *An empirical equation*

The above discussion clearly shows that the size of the spalling defects in a CFRP plate is dependent on various drilling variables. An empirical equation can therefore be obtained by a non-linear regression using all the experimental data. This leads to

$$\ln l = 42.64 - 52.8 \cdot \ln D + 1.97 \cdot \ln v_f - 0.49 \cdot \ln n + 19.68 \cdot (\ln D)^2 + 0.12 \cdot (\ln v_f)^2$$
$$- 1.2 \cdot (\ln n)^2 + 0.85 \cdot \ln D \cdot \ln v_f - 0.87 \cdot \ln v_f \cdot \ln n + 0.28 \cdot \ln D \cdot \ln n$$

where *l* is the spalling size in mm, *D* is the drill diameter in mm, v_f is the feed speed in mm/min and *n* is the drilling speed in rpm.

1.5. Abrasive machining

As we can see in the previous sections, conventional cutting and drilling normally introduce subsurface damages. An effective way to minimize them is to use grinding wheels and abrasive drill bits. For instance, the surface integrity of a ground PMCRLF, as shown in Figure 1.24, is much better compared with that from cutting (Figure 1.10). The material removal mechanisms in grinding and abrasive drilling are more complicated. Some details can be found in [HU 03, HU 04].

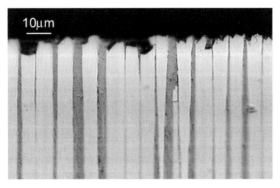

Figure 1.24. *The cross-section of a ground F593 surface (fiber orientation: 90°; depth of cut: 20 μm)*

1.6. Concluding notes

From the above analysis, we can obtain the following conclusions:

– fiber orientation θ is a key factor that determines the surface integrity of a machined PMCRLF component. θ = 90° is a critical angle, beyond which a severe subsurface damage will occur;

– the significant bouncing back of materials after cutting is a characteristic phenomenon associated with the machining of PMCRLF;

– when $\theta \leq 90°$, three distinct deformation zones appear, i.e. chipping, pressing and bouncing. However, if $\theta > 90°$, fiber-bending during cutting will become more significant;

– the rake angle of a cutting tool γ_0 only slightly affects the surface roughness. In the range studied, a better surface will be obtained when $0° < \gamma_0 < 20°$;

– the cure conditions for making the PMCRLF alter their mechanical properties, particularly in the transverse direction. While the cure degree does not influence the surface roughness, it contributes slightly to the change of the cutting forces.

1.7. References

[DAV 03] DAVIM J.P., REIS P., LAPA V., ANTÓNIO C.C., "Machinability study on Polyetheretherketone (PEEK) unreinforced and reinforced (GF30) for applications in structural components", *Composite Structures*, Vol. 62, 2003, p. 67-73.

[HU 03] HU N.S., ZHANG L.C., "A study on the grindability of multidirectional carbon fibre-reinforced plastics", *Journal of Materials Processing Technology*, Vol. 140, 2003, p. 152-156.

[HU 04] HU N.S., ZHANG L.C., "Some observations in grinding unidirectional carbon fibre-reinforced plastics", *Journal of Materials Processing Technology*, Vol. 152, 2004, p. 333-338.

[MAH 01a] MAHDI M., ZHANG L.C., "An adaptive three-dimensional finite element algorithm for the orthogonal cutting of composite materials", *Journal of Materials Processing Technology*, Vol. 113, 2001, p. 368-372.

[MAH 01b] MAHDI M., ZHANG L.C., "A finite element model for the orthogonal cutting of fibre-reinforced composite materials", *Journal of Materials Processing Technology*, Vol. 113, 2001, p. 373-376.

[RAM 97] RAMULU M., "Machining and surface integrity of fibre-reinforced plastic composites", *Sadhana*, Vol. 22, 1997, p. 449-472.

[WAN 03] WANG X.M., ZHANG L.C., "An experimental investigation into the orthogonal cutting of unidirectional fibre reinforced plastics", *International Journal of Machine Tools and Manufacture*, Vol. 43, 2003, p. 1015-1022.

[ZHA 01a] ZHANG H., CHEN W., CHEN D., ZHANG L.C., "Assessment of the exit defects in carbon fibre-reinforced plastic plates caused by drilling", *Key Engineering Materials*, Vol. 196, 2001, p. 43-52.

[ZHA 01b] ZHANG L.C., ZHANG H., WANG X., "A new mechanics model for predicting the forces of cutting unidirectional fibre-reinforced composites", *Machining Science and Technology*, Vol. 5, 2001, p. 293-305.

Chapter 2

Machinability Aspects of Polymer Matrix Composites

This chapter deals with the general aspects of machinability of a polymer matrix, regarding tools, risk of damage and the main parameters involved.

As a direct application of general concepts, the development of a family of router milling tools for the high performance milling of carbon fiber-reinforced plastics (CFRP) is described. The new milling tools are shaped by multiple left-hand and right-hand helical edges, which form small pyramidal edges along the cutting length. Several substrates and coatings were tested. After the analysis of tests and modifications on the tool prototypes, the final result was a series of routing endmills optimized for carbon fiber composites, defining the influence of each milling tool feature on tool performance.

The specific cutting forces, tool wear and others aspects are discussed in depth.

Chapter written by Franck GIROT, Luis Norberto LÓPEZ DE LACALLE, Aitzol LAMIKIZ, Daniel ILIESCU and Mª Esther GUTIÉRREZ.

2.1. The machining of polymer composites

CFRP are widely used in the airframes of commercial airplanes, for example in the case of the Airbus 380 (25%), A350 (52%) and the B787 Dreamliner (50%). Structural parts (wings, keel beam, cowlings, J-nose, etc.) are mainly made of multilayer composite sheets. The components conform to the final shape, but after the composite construction all parts must be trimmed to the correct dimensions and shape, using milling as the main process. Currently other techniques such as abrasive water jet (AWJ) are being studied, but not enough data about the possible material damage is available yet. A limitation of AWJ is when the water jet cuts through one composite face but the jet also damages the unprotected surface part on the other side.

The most widely used composite is glass fibers in polyester resin, which is commonly referred to simply as *fiberglass*. Fiberglass composite is lightweight, corrosion resistant, economical, and it has good mechanical properties. Chemosetting plastics are normally used for GRP production, most often polyester, but vinylester or epoxy are also used. The glass fibers can be in the form of a chopped strand mat or a woven fabric. These composites are used in arrows, bows, helmets, swimming pools, translucent roofing panels, automobile bodies (running boards for example), electrical insulation, machine envelopes and boat hulls.

On the other hand, Kevlar® was introduced by DuPont in the 1970s. Kevlar fibers are highly anisotropic, stronger and stiffer in the axial direction than in the transverse direction. Kevlar is very resistant to impact and abrasion damage. This material can be mixed with carbon fibers in hybrid fabrics to provide damage resistance, increased ultimate strains, and to prevent catastrophic failure modes. Kevlar is used in everything, from airplane parts to reinforced suspension bridge structures and suspension bridge cables, sport materials (kayak, boats), tire wire reinforcements, brakes and clutches in pulp form, not to mention a variety of consumer goods. Kevlar is replacing fiber glass-reinforced plastic in race car bodies and air dams because it does not shatter or leave hazardous debris on the track after a crash.

Figure 2.1. *A trimming operation in a five-axis milling center for composites, by MTorres®*

2.2. Tools

In this section the main tool materials and types are presented, especially for milling of the borders of composite parts. This operation is known as routing, or trimming, as shown in Figure 2.1. Drilling, the other important operation on composite components, is the subject of Chapter 3.

2.2.1. *Tool materials*

Tool materials for the machining of composite parts must meet several specifications. Tools should have good resistance to friction and abrasion at the interface chip/rake face and machined surface/clearance face. An important part in heating the workpiece can cause thermal degradation of the matrix, so the tools must have a good hardness at high temperatures and allow the dissipation of heat generated by friction. The resistance of the cutting edge to abrasion wear is a very important feature because the loss of acuity can cause significant damage in the composite material (fiber pullout, delamination, etc.).

The current tool materials, used industrially, are in increasing order of hardness:

– High-speed steels (HSS): these are high alloy steels containing tungsten, chromium, molybdenum and vanadium. To improve performance, these tools can be coated by physical vapor deposition (PVD) with thin films of hard materials (thickness 1 to 5 µm), mainly TiN, TiCN, TiAlN and AlCrN. They are used occasionally (small series, unimportant quality) due to their very low resistance to abrasion. Piquet [PIQ 99] rejects the HSS tools because their life is too short. Their rapid wear makes fiber shearing difficult, leading to their pullout. Even coated, HSS tools are also excluded because of their low edge acuity. Krishnamurthy *et al.* [KRI 92] indicate that the deformation and peeling of the coating can alter the geometry of the cutting edge resulting in severe damage to the part.

– Carbides (CW): these materials, processed by powder metallurgy, can obtain a high hardness and resistance to abrasion, significantly exceeding HSS materials. This is due to the presence of fine CW, TiC and/or TaC particles. To further improve their performance, they may be coated with hard materials obtained mainly by ionic processes such as PVD or CVD (chemical vapor deposition). A wide variety of coatings is possible (TiN, TiCN, TiAlN, Al2O3, AlCrN, diamond, etc.) with a good adhesion on the carbide tool.

– Cubic boron nitrides (CBN) are presented in various forms: by electrolytic deposition of platelets on a tungsten carbide body, CBN blanks sintered and brazed on carbide plates, or directly sintered on a tungsten carbide substrate.

– Diamond, the hardest material, which comes in many forms: synthetic or natural diamond crystals deposited on a HSS or carbide body with an electrolytic binder (Ni), polycrystalline diamond (PCD) obtained by sintering of diamond grains and a Co binder on a carbide substrate.

Ceramics are not suitable because of their low resistance to mechanical shocks and their susceptibility to thermal shock.

In conclusion, the majority of works on FRP machining recommend the use of carbides or PCD tools. PCD tools are preferred for their better resistance to wear, while carbide tools are chosen for their lower cost.

2.2.1.1. Carbide tools

Piquet [PIQ 99] uses K20 submicron carbide grade. These give the cutting edges an excellent acuity (approximately 7 to 12 µm) while remaining highly resistant to wear. Abrate *et al.* [ABR 92] have also validated the milling of GFRP and CFRP with carbide tools. The use of PCD tools is hampered due to its low toughness, a very high cost and also the difficulty in giving the PCD an appropriate form. The milling of KFRP, for example, requires a geometry of the tool that is often difficult to achieve in PCD [KÖN 84].

Iliescu [ILI 08] points out the influence of carbide grade on the tool wear in orthogonal cutting of CFRP (Figure 2.2). As the wear depends on the loads applied to the tool during cutting, he demonstrates that K20 carbide grade (H13A from Sandvik Coromant®; Table 2.1) performs better than P10 or P40 cermet grades (S1P and S6 from Sandvik Coromant; Table 2.1).

Cermets consist of grades containing a significant proportion of γ-phase, (i.e. TiC, TaC, NbC, etc.) together with WC and Co. The main features of the γ-phase are good thermal stability, resistance to oxidation and high temperature wear. These grades are designed to provide a favorable balance of wear resistance and toughness in applications that generate high temperatures. These conditions arise in cutting.

Figure 2.2 shows that after a machined length *Lc* of 50 m, the mean value of the total cutting load on the S6 and S1P tools is approximately 100 N higher than that of H13A tools. The cutting load increases with the machined length, so the S6 and S1P grade tools wear faster than H13A grade tool.

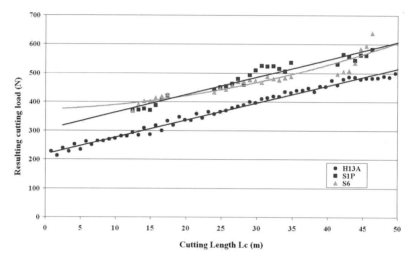

Figure 2.2. *Evolution of the cutting force with machined length Lc for different carbide grades during orthogonal cutting of CRFP (Vc = 60 m/min, rake angle α = 15°, feed rate f = 0.2 mm)*

Carbide or cermet type (by Sandvik®)	Co content (%)	TiC+TaC content (%)	WC content (%)	Grade
H13A	6	0	94	K20 / N15
S1P	10.00	22	68	P10 / S15
S6	11.70	5.50	82.80	P40

Table 2.1. *Chemical composition and grades of carbide and cermet tools*

As pointed out by Figures 2.3 to 2.5, the higher the cobalt content the lower the hardness and the wear resistance, and the higher the toughness. S1P and S6 cermet grades have a similar cobalt level and differ in their titanium and tantalum carbide contents. These carbides have a brittle behavior during machining leading to a lower wear resistance than standard WC-Co grades.

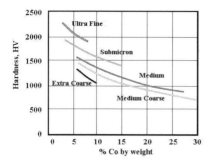
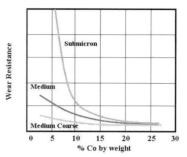

Figure 2.3. *Hardness as a function of cobalt content for various CW grain sizes*

Figure 2.4. *Wear resistance as a function of cobalt content for various CW grain sizes according to ASTM B611-85 test method*

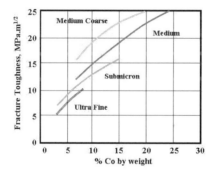

Figure 2.5. *Fracture toughness as a function of cobalt content for various CW grain sizes*

Thus, cermets are traditionally avoided for wear resistance because of being more brittle than standard WC-Co grades. The K20 carbide grade (6% Co and 94% CW) is now up to the more adapted material for composite machining tools.

2.2.1.2. *Coatings for carbide tools*

To improve the performance of the tools, the use of coatings has become very common. The materials used for the latter can be classified into four groups:

1. coatings based on titanium (TiC, TiN, TiB2, TiCN, etc.), often supplemented by aluminum (TiAlN, TiAlON, AlCrN, etc.);

2. ceramic coatings;

3. ultra-hard coatings (diamond);

4. self-lubricant coatings with MoS2 or CW+C as outer layer.

The coatings consist of either a single layer or a superposition of several layers of different materials whose total thickness generally does not exceed 15 µm. Whether of the CVD or PVD type, the deposition process aims to create a close link between the substrate and its coating. The adhesion to the substrate is an essential criterion for the choice of the coating. They will also be selected for their high temperature properties (hardness, toughness, thermal conductivity). For example, the use of alumina Al2O3 in multilayer coatings is dependent on the thermal insulation to retain the mechanical properties of the substrate. The coating is part of the tool in contact with the material workpiece; it will be designed to provide specific properties to achieve given performances. Resistance to wear, friction and chemical inertia are among the characteristics that a coating has to provide for a cutting tool.

For the machining of composite material, the best results are obtained with ultra-hard coatings.

Diamond has ideal properties for machining of composites. On the one hand, its hardness makes it more resistant to abrasive wear than any other cutting materials. On the other hand, its high chemical stability and low affinity with non-ferrous materials prevent the formation of a built-up edge and stabilize the cutting performance at a very high level. One of the advantages of diamond coatings as compared to PCD lies in their adaptability to settle on any geometry. Therefore, tools such as WC carbide mills, drills and inserts can be coated with diamond. The experience of some coating suppliers has showed that the best results are obtained with submicron carbides (0.5 and 3 μm) with a cobalt content below 10% (Figures 2.6 to 2.9).

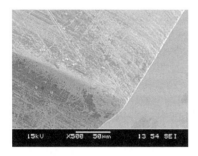

Figure 2.6. *K20 carbide grade tool with good sharpness [ILI 08]*

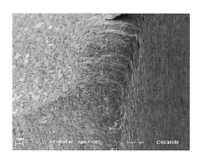

Figure 2.7. *Diamond-coated K20 carbide from Diager® [ILI 08]*

Figure 2.8. *Balzers® diamond-coated K20 carbide tool [ILI 08]*

Figure 2.9. *Cemecon® diamond-coated K20 carbide tool [ILI 08]*

2.2.1.3. *Polycrystalline diamond (PCD)*

Guégan [GUE 94] recommends PCD tools instead of carbides or diamond-coated carbides in order to mitigate the damage associated with cutting. Schulz [SCH 97] studied tool wear in high-speed milling of CFRP. He concluded that only PCD tools provide an economically viable lifetime at cutting speeds around 2,500 m/min. Abrate *et al.* [ABR 92] have also validated the milling of GFRP and CFRP with WC and PCD tools. Ramulu [RAM 99] compared carbide and PCD tools and pointed out that carbide tool wear was about 120 times faster than PCD tool wear. In terms of durability, productivity and repeatability, the PCD tools seem to give better performance. This was noted in several campaigns of comparative tests between carbide tools and PCD tools [CHA 95, HIC 87]. The use of PCD tools is hampered due to its low toughness, a very high cost and also the difficulty in giving the PCD an appropriate form. The milling of KFRP, for example, requires a tool geometry that is often difficult to achieve in PCD [KÖN 84].

Currently, PCD materials consist of a compound of diamond and cobalt particles which is sintered at high temperature (1,600°C) under high pressure (1,000 MPa). The microstructure can be classical (particle mean size around 5, 10 or 25 µm; 92 vol.% diamond and 8 vol.% cobalt) or bimodal (particle size of 4 and 25 µm; 95 vol.% diamond and 5 vol.% cobalt) as illustrated in Figures 2.10 and 2.11. PCD make it possible to obtain tools with a high cutting edge acuity. Their wear behavior depends on the cobalt content and the diamond particle size. The higher the cobalt content or the lower the particle size, the faster the abrasive wear. Their in-use temperature in air has to be below 700°C in order to avoid graphitization of the diamond particles.

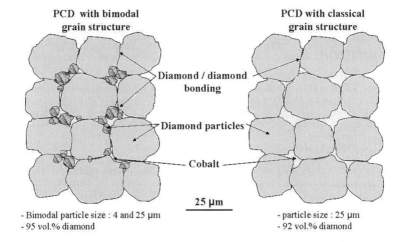

Figure 2.10. *Bimodal and classical PCD grain structure*

Machinability Aspects of Polymer Matrix Composites   49

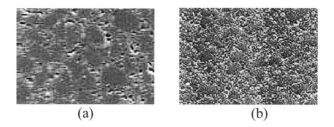

(a)   (b)

Figure 2.11. *Microstructure of bimodal grain (a) and classical (b) PCD from Element Six® (Syndite CTM302 and CTB025 grades)*

2.2.1.4. *Diamond abrasive or diamond grit tools*

These tools are made up of diamond particles deposited onto an HSS or carbide body and embedded in an electrodeposited nickel layer (Figures 2.12 and 2.13). The tool wear behavior depends on the diamond particle size, the diamond particle volume fraction and distribution, the particle embedment, the diamond type. Diamond particles have usually a diameter higher than 125 μm. Their embedment in the nickel layer is between 1/3 and 2/3 of their diameter. The lower the embedment, the longer the tool life, but the more brittle the tool. The best results are obtained with an embedment of 1/2. The higher the volume fraction, the longer the tool life. Better results are obtained with natural diamond than synthetic diamond.

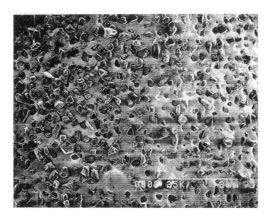

Figure 2.12. *SEM picture of a natural diamond abrasive tool*

Figure 2.13. *Details of the diamond particle embedment in the nickel layer (particle size: 150 μm; embedment. 1/2)*

2.2.2. Tool types

There are two types of tools suitable for routing CFRP materials:

a) with polycrystalline diamond (PCD) inserts; and

b) the solid carbide endmill coated with PVD or with a diamond coating (Figure 2.14).

The latter can be of two types, the helical flute milling tool (somewhat similar to the typical end milling tools for metals) and the type of tool with multiple teeth. As far as the former type is concerned, PCD is an expensive solution, providing long tool life and allowing high feeds and cutting speeds – unfortunately the PCD price/tool life ratio is not adequate for most small and medium enterprises. Tool price is 6 times higher than price of the carbide tools whereas tool life is only 3 times higher. Moreover, the tool geometry of PCD must be simple, with straight flutes of PCD plates brazed on a body made of sintered carbide.

Figure 2.14. *The SGS® routers, made of sintered carbide*

A PCD improvement is vein technology (for example V-tec, by MegaDiamond®, Figure 2.15 or 2.16 by Unimerco), where a combined sintered carbide-PCD rod is ground to the final shape, similar to an helical endmilling tool. The main advantage is the elimination of the brazed joint close to the cutting edge, achieving more tool life than straight flute tools. With this type of tool, cutting speed can be three times the sintered one (for example 18,000 rpm with a 12 mm diameter tool).

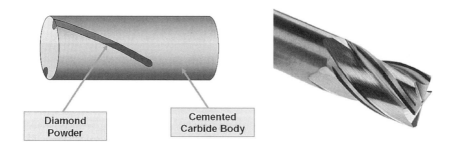

Figure 2.15. *The PCD vein technology, from MegaDiamond Inc®*

However the sintered carbide tools are cheaper and used more by polymer matrix composite machiners.

As far as coatings are concerned, the general PVD types improve the performance of sintered tools. TiAlN grades are suitable for this application. On the other hand, the new diamond coatings applied with CVD are achieving good results (Figure 2.16). Thus, DiaTiger® is made up of interlocking layers of poly and nano-crystalline diamond. The alternating layers provide the strongest protection against abrasive and adhesive wear, and cutting edge damage from mechanical shock.

Figure 2.16. *Four milling tools by Unimerco®. Solid carbide coated by CVD diamond, PCD straight fluted, veined PCD and diamond grit mill*

For the milling, routing and trimming of polymer composites several types of tools are currently on the market, as shown in Figures 2.14, 2.15, 2.16 and 2.17:

– electroplated diamond tools (Figure 2.16 right), in which steel is the binder of mini-grains of diamond attached to the tool shank. The performance of a diamond tool depends on how diamonds are distributed (grit density) and adhered in matrix. Diamond distribution can be random or regular, and its adherence strong or weak. This type of tool is suitable for the trimming of glass and carbon fiber composites;

– the multi-pyramidal-like edge tool, similar to that optimized in section 2.5 and shown in Figure 2.17. They are easy to manufacture using the usual tool grinders in toolmaker's workshops. Controlling the grinding wheels by CNC, the grinding machines produce the tool edges on carbide rods by the right-hand crossing of several helices and several others by left-hand crossing. The final appearance of the tool side surface is a multi-pyramidal-like (knurling-like) surface. The multiple teeth solution eliminates the cutting force along the Z-axis. This fact makes it possible to reduce the workpiece static deflection and vibrations;

– the compression router's unique geometries, in which the flutes intersect from both directions, were developed and refined by an Amamco® (Figure 2.17, center), with a CVD diamond coating. This coating consists of 100% pure diamond crystals that are grown directly on carbide tool blanks in a vacuum chamber using superheated filaments to activate a hydrogen and methane gas process. The resulting vapor mixture bonds to the surface of a carbide tool to a thickness that is controlled within a range of three to 30 microns. In this tool there are two separate sets of left-hand and right-hand flutes that overlap about a third of the way up the router shaft from the tip. The left-hand flutes push the composite down and the overlapped right-hand flutes come past and pull it right back up;

– brazed PCD (Figure 2.16) typically has a hardness in the range of 6,000 Hv. It offers excellent abrasion resistance and very sharp edges. The brazed construction means that PCD can be very economical, especially in bigger diameters;

– the PCD tool called SERF (Sinusoidal Edge Rougher Finisher) by Onsrud® (Figure 2.17) provides the benefits of a roughing mill with the edge quality of a finishing mill in a Polycrystalline Diamond double-flute tool on solid-carbide body.

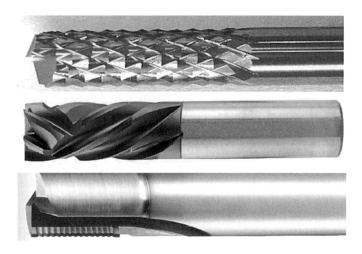

Figure 2.17. *Router tools: (top) the multi pyramidal-like edge tool; (middle) the compression router by Amanco® and (bottom) the PCD SERF by Onsrud®*

2.3. Cutting mechanisms in composite materials

Koplev *et al.* [KOP 80, KOP 83] were the first to look at the cutting mechanisms during composite material machining. The authors have conducted interrupted cutting tests on CFRP for fiber orientations of 0° and 90°. For these tests, a new chip preparation technique has been adopted: adhesive rubber glue is applied on the workpiece surface before machining in order to keep the composite material in the form of "macro-chips". This procedure avoids the powdered form of chips usually encountered in machining of composites. This first attempt was completed in the works of Wang *et al.* [WAN 95], Arola *et al.* [ARO 96], still in orthogonal cutting. All these authors argue that the fiber orientation with respect to the tool displacement is the main factor of influence on the formation of the "chip". They also agree that the nature of the composite (matrix and fibers), the tool rake angle and the edge acuity are major parameters. They distinguish several types of material removal according to the fiber orientation.

2.3.1. Influence of fiber orientation on cutting

For a 0° fiber orientation and a good edge acuity, the mechanism of chip formation consists of a mode I loading (opening) of the cut section with rupture along the fiber/matrix interface (Figures 2.18 and 2.19). The separation of the "chips" then occurs after rupture of the fibers in a direction perpendicular to their axis giving rise to large chips. The crack generated by this cutting mode propagates in front of the tool tip (crack ahead).

In case of a low edge acuity tool, the pressure of the tool on the composite material leads to a buckling of the fibers (Figure 2.18 and 2.20) with rupture along the fiber/matrix interface giving large fragmented debris.

These modes have been confirmed by Iliescu *et al.* [ILI 08] based on high speed video experiments and discrete element simulation. Figures 2.19 and 2.20 point out the contact forces and connection links. The composite parts that are no longer connected to the rest of the workpiece change color (from white to dark gray). In dark gray are shown the compression forces and in light gray the tensile forces, while the existing links are in green whereas others are broken. The pictures given by the discrete element simulation highlight the buckling and delamination of the fibers.

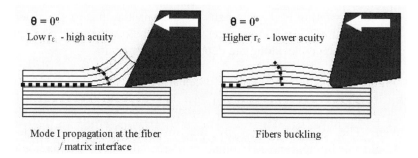

Figure 2.18. *Mechanisms of "chip" formation during the machining of 0° unidirectional FRP with high or low cutting edge acuity tools*

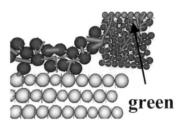

Figure 2.19. *DEM simulation of the cutting of 0° unidirectional FRP pointing out the mode I opening*

Figure 2.20. *DEM simulation of the cutting of 0° unidirectional FRP pointing out the buckling of the fibers and their rupture*

For a fiber orientation around 90°, the material removal is initiated by a mode I opening, which penetrates the material (under the direction of the cut) following the fiber/matrix interface, extended by a secondary rupture that goes back to the surface due to the fibers shearing (Figure 2.21). The compression zone below the tool results in significant long cracks along the depth of the composite.

Figure 2.22 shows the contact forces and connection links. In dark gray are shown the compression forces and in light gray the tensile forces, while the existing links are in green and the purple ones are broken. The pictures given by the discrete element simulation highlight the flexion and delamination of the fibers. This behavior is in agreement with the observations of Koplev [KOP 83] who pointed out that the compression zone below the tool led to the apparition of cracks of 0.1 to 0.3 mm in the material depth for that fiber orientation. When a low edge acuity tool is used these phenomena are exacerbated.

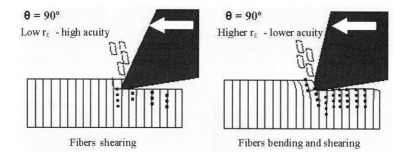

Figure 2.21. *Mechanisms of "chip" formation during the machining of 90° unidirectional FRP with high or low cutting edge acuity tools*

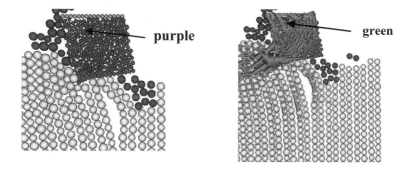

Figure 2.22. *DEM simulation of the cutting of 90° unidirectional FRP pointing out the penetrating cracks and the fibers shearing*

For fiber orientations of 45°, the cutting mechanism consists of a tensile deformation of the fibers and a shearing by the cutting edge perpendicular to the fibers. The chip is then formed by shearing along the fiber/matrix interface to the free surface (Figures 2.23 and 2.24). Then the cut fibers undergo a springback to lie in a plane above the theoretical machined surface. The fibers rub against the tool clearance face, causing abrasive wear of the latter. Figure 2.24 shows the contact forces and connection links. In dark gray are shown the compression forces and in light gray the tensile forces, while the existing links are in green and the red ones are broken. The pictures given by the discrete element simulation highlight the tensile deformation and the shearing of the fibers. For this orientation of the fibers, subsurface delamination of the material has been pointed out. When a low edge acuity tool is used these phenomena are exacerbated.

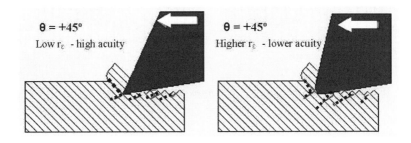

Figure 2.23. *Mechanisms of "chip" formation during the machining of +45° unidirectional FRP with high or low cutting edge acuity tools*

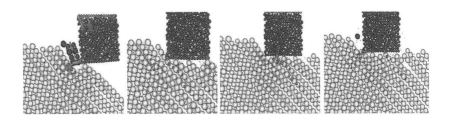

Figure 2.24. *DEM simulation of the cutting of +45° unidirectional FRP pointing out the cutting mechanisms and the rubbing of the cut fibers on the tool clearance face*

During the cutting of fibers oriented at -45°, fiber bundles are pushed by the tool, bent and broken by bending (Figures 2.25 and 2.26).

Cracking occurs in the thickness of the material at the fiber/matrix interface of these bundles. Discrete element simulation results show these phenomena more explicitly. Noticeable defects propagate into the part. The resulting chips are large bundles of matrix and fibers. The cut fibers are then positioned in a plane above the theoretical machined surface. There is less rubbing of the fibers leading to lower abrasive wear of the clearance tool face.

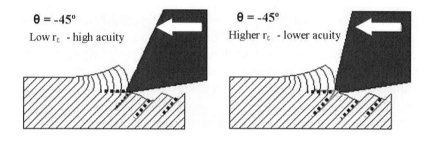

Figure 2.25. *Mechanisms of "chip" formation during the machining of -45° unidirectional FRP with high or low cutting edge acuity tools*

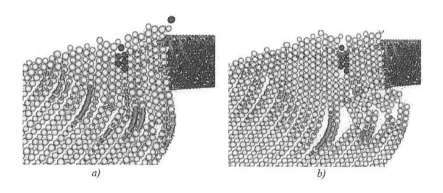

Figure 2.26. *DEM simulation of the cutting of -45° unidirectional FRP pointing out the flexion of the fiber bundles and the resulting multicrack damage*

2.3.2. Influence of fiber orientation on tool wear

The abrasive wear of the tool material is mainly due to the friction at the chip/tool and tool/workpiece interfaces by hard constituents belonging to the machined material (fibers mainly).

The abrasive wear is characterized by the formation of ribbed zones or by the loss of sharpness of the cutting edge (increase of the cutting edge radius). As the cutting edge radius is of the same magnitude as the fiber diameter, fibers are cut cleanly. As soon as the cutting edge radius is approximately five to ten times the fiber diameter, fibers are repulsed and cut badly.

The abrasive wear appears on both the rake and clearance faces of the tool, however this wear is higher on the clearance face. The fiber debris and the friction of embedded fibers are responsible for this wear. The low cutting edge acuity of the tool creates repulsion of workpiece material and higher friction (Figure 2.27) [GHI 03; ILI 08]. Combined with a low speed rate of the tool, this friction between the composite material and the clearance face generates heat at the workpiece surface.

The cutting mechanisms previously described indicate a more intense friction of the fibers on the clearance face of the tool for a fiber orientation at +45 °, and to a lesser effect for orientation at 0° and 90 °.

This effect is amplified when the cutting edge sharpness of the tool decreases (Figure 2.27). The quantity of material which is repulsed by the tool is then higher, increasing friction and therefore wear.

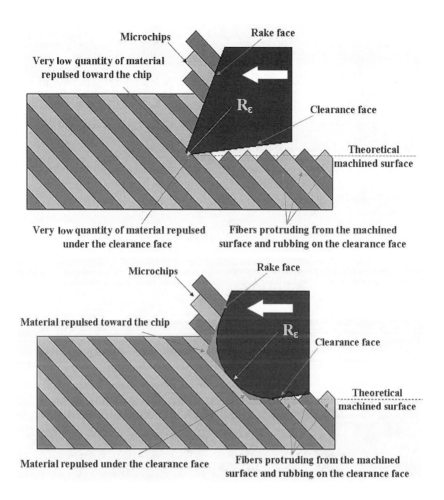

Figure 2.27. *Influence of cutting edge acuity on the fiber friction onto the clearance face for a +45° fiber orientation*

This effect has been clearly identified by Iliescu *et al.* [ILI 08] in orthogonal cutting tests of a multilayer CFRP (Figure 2.28). The tool then presents wear faces of the wave-shaped type. The difference in abrasive wear between the different orientations of the layers is evident.

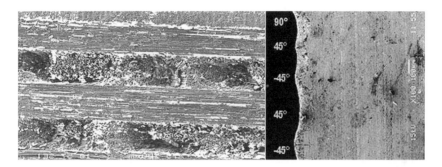

Figure 2.28. *Tool abrasive wear behavior as a function of CFRP layer orientation after a cutting length Lc=100 m (K20 grade carbide tool; rake angle α= 0°; cutting speed Vc=60 m/min; feed rate f=0.1 mm/rev)*

This behavior has also been reported by Girot [GIR 97] when CFRP or CFRC are milled with diamond abrasive tools (Figure 2.29).

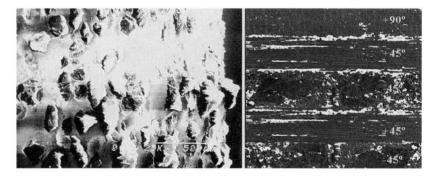

Figure 2.29. *Diamond abrasive tool behavior as a function of CFRP layer orientation after a cutting length Lc=3000 m (diamond particle size 250 μm; cutting speed Vc=900 m/min; feed rate f=0.05 mm/rev)*

2.3.3. Influence of fiber orientation on cutting loads

During orthogonal cutting of unidirectional CFRP, cutting and feed loads Fc and Fh have been measured using a piezoelectric dynamometer [ILI 08].

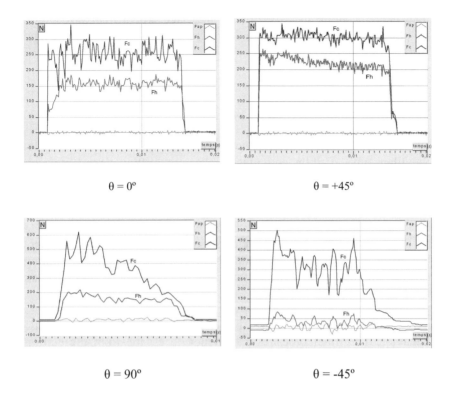

Figure 2.30. *Cutting and feed loads during orthogonal machining of CFRP (Vc=60 /min, feed rate f = 0.2 mm and rake angle α = 0°)*

Figure 2.30 shows that the feed load *Fh* corresponding to the friction of the composite material on the clearance face decreases when the fiber orientation evolves from +45°, 90°, 0° to -45°. The cutting load *Fc* is almost constant for the +45° orientation when it is cyclical for the other orientations.

For the 0° orientation, the cutting load *Fc* evolves cyclically between a maximum and minimum value corresponding to the buckling of the fibers and the propagation of a crack in the cutting direction at the fiber/matrix interface. For a fiber orientation of 90°, this evolution is due to the fiber cutting by shearing and the propagation of a crack in the direction perpendicular to the cutting displacement along the fiber/matrix interface. For a fiber orientation of -45°, the cyclical evolution is mainly due to the repulsing and cutting of the fibers by flexion and a crack propagation along the fiber/matrix interface.

2.3.4. *Influence of fiber nature on cutting*

The cutting mechanisms of composite materials depend greatly on the fiber type. Carbon and glass fibers are brittle components and although they have different microstructures (Figure 2.31) they are very sensitive to the notch effect produced by the tool tip. They are always cut by the shearing produced by the tool.

Glass fibers have an abrasive effect on the tool when carbon fibers are difficult to cut because of their high Young's modulus depending on the graphitization level of the fiber. Carbon fibers also have an abrasive effect on the tool.

Aramid fibers are relatively hard organic materials with a ductile behavior. This fiber type has a tendency to escape from the cutting edge and to lint when cutting. To produce a clean cut, it is necessary to pull the aramid fiber in order to shear it easily with a well sharpened tool.

The ductile rupture of the aramid fiber is the result of its macromolecular structure which has weak molecular bonds in the radial direction which allows slippage between the molecular chains. This explains the ease of the fiber to lint. Its low resistance to compression facilitates the contraction of the fiber in the matrix and its tendency to escape from the shear.

This peculiarity of the aramid fiber means that tools with special and adapted geometries are necessary to machine KFRP materials.

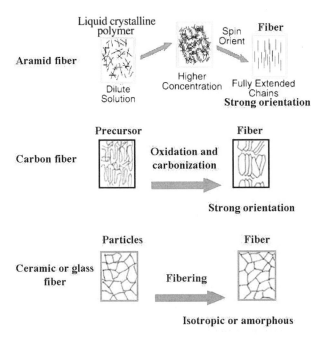

Figure 2.31. *Different microstructures resulting from the manufacturing process of current industrial fibers*

2.4. Composite material damage due to machining

The machining of composite structures is quite different from the machining of metallic parts and can cause several defect types. Many works have pointed out these defects [INO 97, PIQ 99, HO-C 92, GUE 94, KÖN 89]. Most problems encountered in the composites are related to the quality of machining; the most relevant damage being classified as follows:

– mechanical damage: fiber linting, delamination, fiber pullout from the machined surface, generation of multiple cracks on the material surface and others;

– thermal damage: burn of the matrix, etc.;

– chemical damage: water recovery of the matrix and destruction of the links at the fiber/matrix interfaces.

The mechanical and thermal loading of the composite material during the machining operations generate these damages.

2.4.1. *Mechanical damage*

2.4.1.1. *Fiber linting*

Linting is characterized by a poor fiber cutting by the tool. The cut fibers have a frayed aspect with a flaking of the surface during drilling or milling operations. This is mainly present during the cutting of aramid fiber composites (Figure 2.32). It is mainly associated with an incorrect definition of the tool geometry and unoptimized machining conditions.

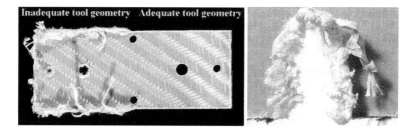

Figure 2.32. *Linting phenomenon which occurs with aramid fibers when using an inadequate tool geometry*

2.4.1.2. *Fiber pullout and matrix decohesion*

Fiber decohesion is characterized by the detachment of unbroken fibers from the matrix. The flexion of the fibers under the influence of the cutting edge on the machined surface leads to a relative displacement of the fibers with respect to the matrix, similar to a slippage causing decohesion of the fiber/matrix interface.

The pullout occurs when the fiber orientation with respect to the tool displacement is between -15° and -75°. Fibers are subject to stresses which can cause the tearing of fiber and matrix pieces, leaving cavities on the machined surface (Figure 2.33). This is more present in the case of highly oriented laminates or when the tool sharpness decreases. This phenomenon is repeated at each ply having the same angular orientation between the tool displacement and the fibers. No model has been developed so far to reflect the occurrence of this phenomenon.

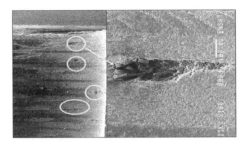

Figure 2.33. *Fiber pullout phenomenon repeated for the same ply orientation and details of the defect*

2.4.1.3. *Delamination*

The mechanism of delamination is characterized by the separation of the plies in the thickness of the composite and by the formation of interlaminar cracks in the material. This damage occurs in drilling, particularly when the drill begins to penetrate the composite material (lifting of the plies in the periphery of the drill due mainly to the rake and the helix angle of the drill) and/or when the drill exits the hole (pushing on the last plies by the drill). In milling, this is characteristic of the geometry of a tool which separates the superficial plies of the central layers (Figure 2.34).

Intralaminar cracks can also be produced by the machining as mentioned in section 2.3.1, mainly due to the orientation of the fiber layers with respect to the tool displacement as illustrated in Figure 2.35.

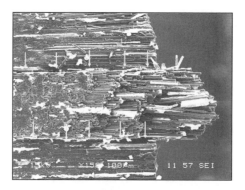

Figure 2.34. *Delamination between the first layers of a CFRP material due to milling*

Figure 2.35. *Penetrating cracks due to the orientation of the fibers with respect to the tool displacement (90°) and revealed by a liquid penetrating technique*

2.4.2. Thermal damage

An excessive release of heat during the machining may lead to a risk of thermal degradation. Overall, this risk causes carbonization of thermosetting matrices and fusion of thermoplastics matrices. It can also damage the fiber (burning of the carbon fibers). This damage occurs routinely in binary form: presence or absence of thermal degradation.

The thermal damage is mainly due to:

– an inadequate or non-existent evacuation of the heat generated by cutting associated with an inappropriate choice of election of the tool material (dry drilling of a GFRP with a carbide tool for example; see Figure 2.36);

– an excessive cutting speed (dry milling of CFRP; see Figure 2.37).

This phenomenon is easily controllable providing an adequate selection of the tool material, an appropriate cooling device and a control of the rotation speed of the tool.

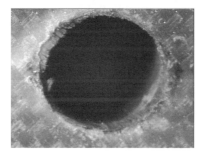

Figure 2.36. *Burning of the epoxy resin during dry drilling of GFRP with a carbide tool (cutting speed Vc= 300 m/min)*

Figure 2.37. *Burning of the CFRP epoxy matrix, due to the bad cutting of worn tools, after 42 mm of the slotting operation (see section 2.5.2)*

2.4.3. *Chemical damage*

Chemical damage can occur when machining is performed under lubrication and depends on:

– the lubricant or cooling fluid type (water-based or oil);

– the lubricant or cooling fluid characteristics (acidic, basic or neutral pH) which can react with the fibers;

– the contact time with water when used as cooling fluid. There may be a water recovery by the composite resulting in the destruction of molecular bonds between the fiber and the matrix.

This usually leads to a decohesion at the fiber/matrix interface that can generate cracks or delaminations. Dry machining, micro-lubrication or gas cooling help to avoid such damage.

2.5. Milling of composite materials

2.5.1. *GFRP routing*

2.5.1.1. *Tool geometries*

Routing is an often used process for the finishing of GFRP material parts in the automobile, railway and naval industries. There are three types of the most classical tools which enable quality machining with acceptable lifetime:

– multiple teeth routers with an upcut/downcut diamond design that effectively shear fibrous materials;

– diamond grit or diamond abrasive routers;

– PCD routers.

For operations requiring a fine finish, multiple teeth diamond-cut routers with more than 15 crosscuts or diamond grit routers are enough. Otherwise, PCD routers allow a very fine finishing due to the sharpness of their cutting edges. They shave the fibers cleanly and give a perfect edge.

K20 carbide grade multiple diamond-cut teeth router	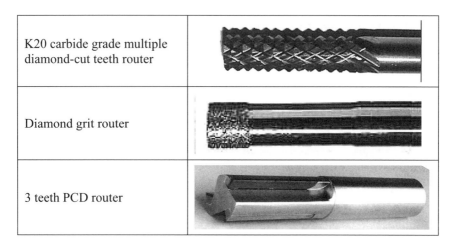
Diamond grit router	
3 teeth PCD router	

Figure 2.38. *Commonly used fiberglass routers*

2.5.1.2. *Routing conditions*

The usual routing conditions are given in Table 2.2. Diamond grit and PCD routers need higher cutting speeds than carbide routers. Whatever the router, the feed rate is usually around 0.05 mm/tooth/rev. A higher feed rate enables us to favor mechanical damage (delamination, multiple cracking).

	Cutting speed Vc, m/min	Feed rate f mm/tooth/rev
K20 carbide grade multiple teeth router	100	0.05
Diamond abrasive router	900-1500	0.05
PCD router	900-1500	0.05

Table 2.2. *Routing conditions for GFRP parts when using carbide or diamond tools*

2.5.1.3. Surface quality and tool lifetime

Figures 2.39 and 2.40 point out a strong correlation between cutting and feed loads and mean roughness Ra when the GFRP composite is milled with a PCD tool.

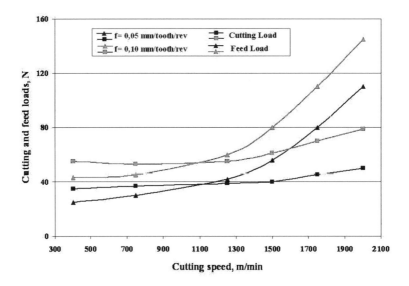

Figure 2.39. *Evolution of cutting and feed loads with cutting speed for a GFRP composite milled with a PCD tool (feed rate: 0.05 mm/tooth/rev)*

The higher the cutting speed, the lower the cutting and feed loads and the lower the mean roughness Ra. However, for cutting speeds that are too high (over 1,300 m/min), the mean roughness increases again corresponding to a strong increase of cutting and feed loads. In that case, the optimum range value of cutting speed seems to be between 1,100 and 1,300 m/min.

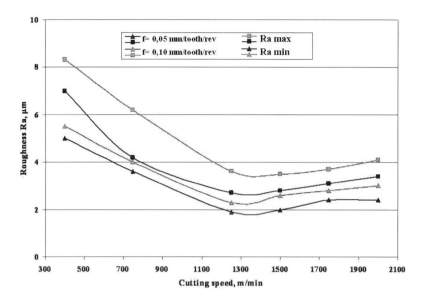

Figure 2.40. *Evolution of roughness Ra with cutting speed for a GFRP composite milled with a PCD tool (feed rate: 0.05 mm/tooth/rev)*

The tool wear depends strongly on the tool material and on the cutting edge characteristics. Figure 2.41 highlights the wear behavior of a K20 carbide grade multiple teeth router and 2 PCD routers. The carbide tool is made of medium size grains (2 μm) with an edge radius of the pyramid tooth of around 10 μm. The PCD routers are made of diamond particles with diameters 10 and 2 μm respectively leading to edge radii of 8 and 5 μm.

The carbide tool wears rapidly in comparison with the two PCD tools. However, in the case of PCD, the lower the particle size, the sharper the cutting edge and the smaller the wear rate.

The evolution of the maximum roughness Rt shown in Figure 2.42 is similar. As the carbide tool has an edge radius higher than the PCD edge radius, roughness Rt is higher. There is a strong correlation between the edge radius and roughness Rt. The smaller the edge radius, the smaller roughness Rt.

A sharp cutting edge is then a condition for a longer tool life and a higher surface quality (low roughness) when milling GFRP composites.

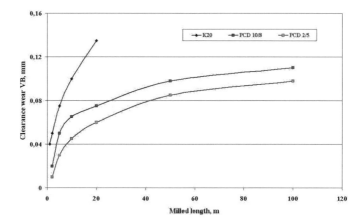

Figure 2.41. *Evolution of tool wear with machined length when milling GFRP composite with K20 carbide grade multiple teeth tool (grain size: 2 µm – edge radius: 10 µm) and, PCD 10/8 (grain size: 10 µm – edge radius: 8 µm) and 2/5 (grain size: 2 µm – edge radius: 5 µm) tools*

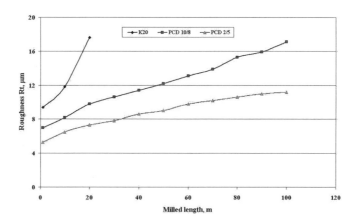

Figure 2.42. *Evolution of roughness Rt with machined length when milling GFRP composite with K20 carbide grade multiple teeth tool (grain size: 2 µm – edge radius: 10 µm) and, PCD 10/8 (grain size: 10 µm – edge radius: 8 µm) and 2/5 (grain size: 2 µm – edge radius: 5 µm) tools*

2.5.2. *Milling of CFRP*

As an application of the general concepts discussed above, the design and testing of a new family of multiple teeth edge routers are presented [LOP 09]. The reader must bear in mind that all parts made of composites must be trimmed to their final shape.

Figure 2.43. *Routing operations on CFRP to evaluate cutting forces*

Five series of experiments were performed, using different tools for each goal. In all the cases, toolholders were high-power shanks with a cylindrical collet, often used in industrial applications (Figure 2.43). Thus, the tests were the following:

– *T0*: tests to define the specific cutting force, in order to compare different CF composites between them and with respect to other well-known materials (steels, aluminum alloys, etc.). For this purpose, helical end milling tools were used, coated by Diamond Like Carbon (DLC). Using this type of endmill sinusoidal cutting forces were recorded and therefore simple calculations led us to obtain the specific cutting force values.

– *T1*: tests to define the most adequate carbide grade for the tool substrate. The same multiple teeth geometry was used but with different grades of the carbide substrate; a new coating based on a nanostructure TiAlN-SiC was also tested. These were the first trials and in view of the results, the study of each tool and process aspects (geometry, substrate, coating and cutting parameters) separately was recommended.

– *T2*: test series to define the best coating or coatings, taking into account the results of the previous *T1*. Along this series, tools with the same multiple teeth geometry were used, but with different coatings.

– *T3*: test to define the best pyramidal-like edge geometry, with a different number of right-hand and left-hand flutes. The number of helices ground around both rotational directions on the carbide rod determines the pyramidal-like formation and therefore the aspect of cutting edges.

– *T4*: testing of a diamond coating applied with a CVD technique, to be compared with the AlTiN coatings.

Figure 2.44. *Tools for tests T0 and T1-T4*

A special comment about the aspiration of dust must be made. The equipment used in the test is the most powerful available, 4 kW, with a maximum depression of 4.600 daPa. A special hopper to obtain all the dust produced is located under the machining zone.

Two composites were tested, both of them multidirectional fiber laminated; the first one was "epoxy matrix with carbon fibers" and the second "epoxy with carbon and kevlar fibers". The latter presents a higher strength and modulus than the former due to the kevlar fibers.

Carbon-aramid hybrid constructions present the high strength and stiffness of carbon and simultaneously the impact protection of aramid. The two composites tested are described in Table 2.3.

The multitooth tool geometry at the tool tip presents two straight edges (Figure 2.44) for the insertion of the tool in the composite at the start of trimming. Damage in this zone was always light, with no influence on tool life.

Figure 2.45. *Aspect of the tool tip, showing some damage after several meters of machining*

Composite 1	
Carbon fiber, resin de Epoxy cured to a 180°C (K3000)	
Pre-impregnated W3T-182-42-F263-27	
Plain weave	
Tensile strength: 525 MPa	Tensile modulus: 53 - 69 GPa
Composite 2	
Matrix Epoxy phenol (10% - 30%)	
Carbon fibers 7782-42-5 (50%)	
Glass fibers	
Kevlar fibers 26125-61-1 (Poly(terephthaloylchloride/p-phenylenediamine))	
Acetone (<2%), Aniline 5026-74-4 (10% - 30%), Derivate of aniline (10% - 30%)	
Plain weave	
Tensile strength: 2,690 MPa	Tensile modulus: 165 GPa

Table 2.3. *Composition and mechanical properties of the tested composites*

2.5.2.1. *Criterion for the wear measurement*

Tool wear has been measured with a special criterion defined by the authors because this type of tool with multiple small edges is not considered in the ISO standards for tool life testing. The wear area was rhombic and the wear zone diagonals were measured using an optical microscope equipped with micrometers for the measurement. The area of the worn surface is calculated as:

$$\text{Worn surface} = (a * b) / 2 \qquad [2.1]$$

where a and b are the diagonals of the wear surface, as shown in Figure 2.46. The percentage of the worn area with respect to the area of the pyramid base is

$$\%\text{Wear} = (\text{Worn surface}/\text{Area of the pyramid base}) * 100 \qquad [2.2]$$

This percentage was a good indicator of the tool degradation. The machining operation was slotting, where the axial depth of cut a_p matches with the composite thickness.

(a) (b)

Figure 2.46. *(a) Pyramidal edge at the initial state previous to cutting; (b) worn surface on top of the pyramid due to the abrasion of the milling process*

There is not a common criterion to consider if a tested tool is too damaged. Users can detect that a bad cutting is being produced by the adherence of resins and fibers to cutting edges and in some occasions because composite parts start to burn. At the same time a higher sound and vibration is produced; the higher the wear the higher the sound. In some occasions there is a sudden tool breakage, located at the interface of the tool with the toolholder.

In our work, during all tests the machining sound and adherence were checked by the direct observation of operator. This is a subjective factor, but a physical measurement of tool wear is always carried out after all tests. Sound emission depends a lot on the part's main geometry and features and on the overhang of each component detail; therefore it is not a correct indicator for a quantitative measurement of wear in industrial applications.

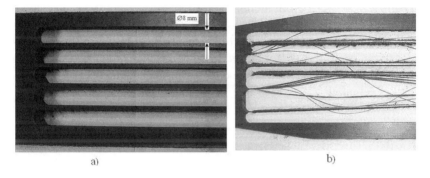

Figure 2.47. *a) View of a good finish with a new tool; b) view of Composite 2 after 30 m of machined length*

In Figure 2.47 the final view of a test part after trimming is shown, with a good appearance (without delamination, thermal damage of grooves). The layers of fiber reinforcements are evident. It is impossible to carry out a roughness measurement due to the soft behavior of the composite under the action of the roughness meter stylus.

At the end of tests, a burnt area around the trimmed line always appears on composites. Damage of the material must always be prevented and therefore this is another signal to detect a worn tool: the first indication of bad appearance along the cut line prompted the end of the test.

2.5.2.2. Cutting force and "specific cutting force"

In [RAH 99] and [RAO 08] cutting forces were analyzed for several CFRP types, proposing two models for the cutting force calculation. In [JAN 00] an experimental method using turning tried to simulate the effect of the router mills in the peripheral trimming of a composite part. However, turning is a continuous operation and milling an interrupted one, so it can affect the comparison.

Several tests (Table 2.4) were performed with a 10° helical $\varnothing 10$ mm z 4 endmill (by Mitsubishi®), measuring the cutting force by means of the Kistler dynamometer plate (type 99255B). The axial depth of cut was the thickness of each part piece; therefore, this was input data, different for each test because test pieces were remnants of real parts, each with different thickness. This aspect causes dispersion in results, surely due to the presence of more kevlar layers inside the thicker remnants.

After recording the forces, both the maximum X and Y force components were divided by the chip section, (a_p*fz), to obtain the basic specific coefficients (equation [2.3]), i.e. the *cutting coefficient* (Ks_x from the X measurement) and the *feed coefficient* (Ks_y from the Y measurement). Maximum rotation speed in tests was restricted to 2,000 rpm by the low natural frequency of the Kistler plate; the tooth passing frequency with an endmilling tool is $N*z$ and the Kistler frequency approximately 1 kHz.

$$Ks_x = \frac{F_{tooth\ X}}{a_p \cdot fz}$$

$$Ks_y = \frac{F_{tooth\ Y}}{a_p \cdot fz}$$

[2.3]

N[rpm]	F[mm/min]	D[mm]	Vc[m/min]	f[mm/rev]
2,000	50	10	62.8	0.025
2,000	60	10	62.8	0.03
2,000	70	10	62.8	0.035
2,000	80	10	62.8	0.04
2,000	90	10	62.8	0.045
2,000	100	10	62.8	0.05
2,000	110	10	62.8	0.055
2,000	120	10	62.8	0.06

Table 2.4. *Cutting parameters for force tests*

Figure 2.48 shows the results, with more test points in the case of composite 2 because the spread of results were wider as it was more difficult to machine. Two main conclusions can be pointed out:

– The specific cutting force is about 300-600 N/mm^2, much lower than that of steel (2,000-2,200) or aluminum wrought alloys (700-900). This fact suggests that tool wear is not caused by a high cutting force, as will be shown in following sections, but by the high abrasion suffered by the tool edges.

– Hybrid composites with both carbon and kevlar fibers present a specific cutting force that is 20-30% higher, due to the higher strength of kevlar fibers with respect to carbon ones. For this reason the more difficult-to-cut composite 2 was used to optimize the milling tools, the main objective of this work.

Results do not match exactly with [RAH 99] where specific cutting forces of 370 N/mm^2 were deduced from the values of cutting forces obtained in turning tests, and with [RAO 08] where values of 50-400 N/mm^2 were calculated depending on the relative orientation of cutting speed with respect to fiber direction. On the other hand, in [RAH 99] cutting forces were higher in the radial direction than in the cutting direction. The reason for the difference is the multidirectional laminated plain weave of CFRP machined in the current work. Each composite is different from the others; therefore, the specific cutting force obtained in this research work can only be considered a qualitative reference.

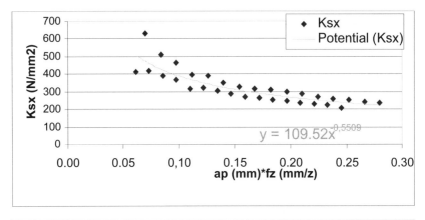

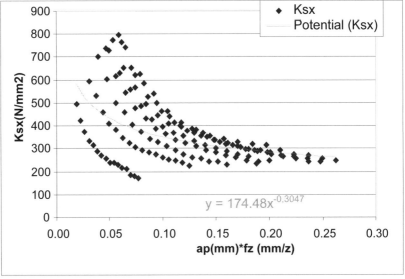

Figure 2.48. *Results of specific cutting force for the CF (composite 1) and CF+KF (composite 2), fitted by exponential laws*

The performance of the helical endmilling tools (T0 series) was very good, but forces along the Z axis were also recorded (the helix angle was 10°), which is not good for flexible parts. The fixing of composite components is very complex and the stiffness of parts along the tool axis is generally very low. For this reason the multiple teeth milling tool was considered as a better option, because no Z-axis force component is produced. In view of the results, a model for the calculus of cutting forces is proposed, taking into account the effect of chip thickness (a_p*fz), introducing a somewhat similar approach to the so-called "size effect" of the cutting of metals:

$$F_t = K_t \cdot a_P \cdot f_z^p$$
$$F_r = K_r \cdot a_P \cdot f_z^q \qquad [2.4]$$

where F_t is the tangential or cutting force (in the Vc direction), F_r is the radial force, K_t is the "specific cutting force", K_r is the "radial specific cutting force", p and q are exponents affecting the feed per tooth fz.

For composite 1:

$$F_r = 55.17 \cdot a_P \cdot f_z^{0.232}$$
$$F_t = 138.271 \cdot a_P \cdot f_z^{0.506}$$

For composite 2:

$$F_r = 103.057 \cdot a_P \cdot f_z^{0.407}$$
$$F_t = 125.945 \cdot a_P \cdot f_z^{0.629}$$

2.5.2.3. Test of different carbide substrates

After the force models and the selection of the multiple teeth endmilling tool type (provided by Kendu®), the following objective in the investigation was to define a good substrate (carbide grade). Three different grades were tested (Table 2.5).

Milling Tool	Carbide grade	Coating
D12x32x90	6% Co medium	No coating
D12x32x90	6% Co medium	naCo 2.7 µm
D12x32x90 D35T	8% Co ultra fine	No coating
D12x32x90 D35T	8% Co ultra fine	naCo 2.7 µm,
D12x32x90 T	12% Co submicron	No coating
D12x32x90 T	12% Co submicron	naCo 2.7 µm

Table 2.5. *Tools tested to define the best carbide substrate*

At the same time there was the opportunity to study the feasibility of using a new nanostructured type of coating, so-called naCo, i.e. nanocomposite (nc-AlTiN)/(α-Si3N4), applied with Platit® Technology by Metal Estalki®.

By depositing different types of materials, the components (like Ti, Cr, Al, and Si) are not mixed and two phases are created. The nanocrystalline AlTiN or AlCrN-grains become embedded in an amorphous Si_3N_4 honeycomb matrix. This nanocomposite structure improves the physical characteristics of the coating layer significantly.

Wear was measured along machining, resulting in the data shown in Figure 2.49, in this case for the 6% Co medium grain with and without coating. In all cases the evolution was similar to this case, with the continuous degradation at the top of pyramidal edges. In view of Figure 2.49, only a slight effect of the naCo coating was detected. The pictures of the wear zone included in this figure show that tool wear grows gradually, affecting the top of pyramidal edges. Coating seems to be a weak barrier against the high abrasion suffered by the pyramid core after the top is damaged during the initial meters of milling.

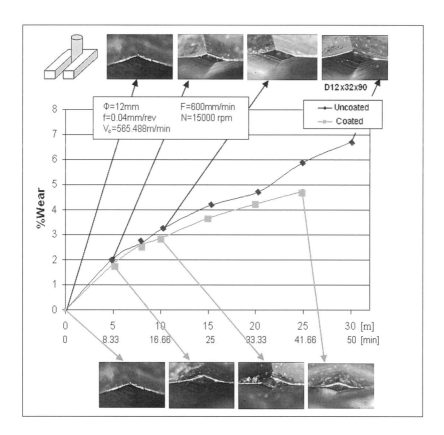

Figure 2.49. *Results for the 6% Co micrograin milling tools, coated and uncoated*

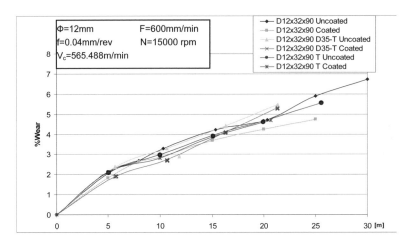

Figure 2.50. *Results for all carbide substrates of Table 2.3, coated and uncoated*

In view of the results for all the tested tools in Figure 2.50, the medium grade was considered the best for the application, with no better results for the submicron and ultra-fine grades. On the other hand, the naCO did not seem very interesting because a much longer tool life was not achieved. For this reason the use of monolayer AlTiN or AlCrN coatings was considered the most feasible option to be applied in the following tests.

2.5.2.4. *Test of coatings*

The testing procedure *T2* was performed to define the best AlTiN or AlCrN coatings. In Table 2.6 the tested tools are shown, in this case 8 mm diameter, made with submicron grain substrate. Tools with this diameter are cheaper than those previously used with 12 mm. Cutting speeds were the same as in the previous tests.

Results are shown in Figure 2.51, for all cases. The best results correspond to the TiAlN+ coating case. This fact could be explained by the high abrasion of the wear process: the thicker layer of this coating (more than 4 microns) resists against abrasion over more time and length than the other tested coatings. The thicker coating layer delays the moment when the top of pyramids lose the protection of the coating and the carbide of the edge core is exposed to abrasion. In addition, when the carbide core is exposed the thicker coating of the lateral borders of the rhombic wear area also acts as a barrier against abrasion.

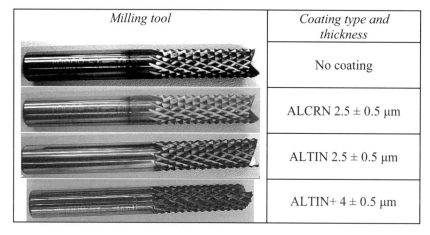

Milling tool	Coating type and thickness
	No coating
	ALCRN 2.5 ± 0.5 µm
	ALTIN 2.5 ± 0.5 µm
	ALTIN+ 4 ± 0.5 µm

Table 2.6. *Tools tested to define the best coating*

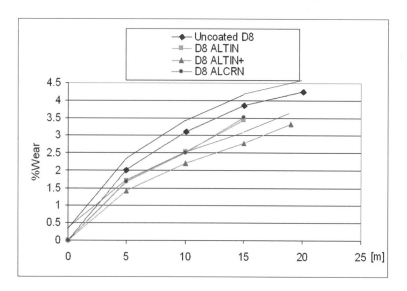

Figure 2.51. *Tool wear for the tests of Table 2.4*

2.5.2.5. Tool geometry test

After carbide and coating selection, two tool geometries were tested, one with 14 right-hand helices and 12 left-hand helices (z14-12), and the other with 14 right-hand helices and 11 left-hand (z14-11), both of which are shown in Figure 2.52. Results were similar, and no effect on wear or test part quality was observed. Perhaps in CFRP delamination this aspect may have some influence but further work is required. When both tool types were new there was good finishing on the milled surfaces, and when tools suffered a severe wear after 25 m of cutting length a bad cutting was suffered in both cases.

With respect to tool wear, tool *z14-11* shows a slightly longer life than the other one, reaching more than 30 m. After a machined length of 20 m, a lot of fibers are adhered on the *Z14-12* edges, causing a slightly poorer cutting process.

Figure 2.52. *a) Milling tool z14-12; b) design of milling tool z14-11*

2.5.2.6. Test of tool with diamond coating

A final test was performed using a CVD diamond coating, applying the conditions marked in Figure 2.53. The first slotting meters showed a damage at the pyramid tops but after 5 m of routing the edge wear was stabilized. In fact, this tool achieved more meters than AlTiN+ with a good cutting, with a very regular wear land at the top of edges.

The cost of diamond coatings is three times that of AlTiN and so the results are under discussion for an economical recommendation of this type of coating.

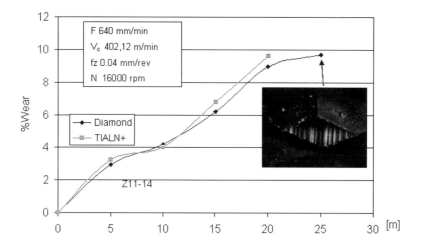

Figure 2.53. *Milling tool z14-11 with a diamond coating versus AlTiN+*

2.5.3. KFRP routing

2.5.3.1. Tool geometries

As discussed previously, special routers are necessary for milling KFRP material parts and usually this type of router is known as a compression router. Due to its high tensile strength, KFRP composites are difficult to cut. It is necessary to pre-load the fibers under tension while simultaneously cutting them in a shearing action.

A compression router has separate sets of left- and right-hand flutes that overlap. The left-hand flutes push the fiber layers down and the overlapped right-hand flutes come past and pull the fiber layers back up. It shaves the fibers cleanly and gives a perfect edge (Figure 2.54).

To minimize vibrations, it is better to support or clamp the composite as close as possible to the cutting tool. Vibrations can cause the resin to break ahead of the cutting tool, which in turn lessens the tension on the fibers and allows them to be pulled instead of cleanly cut or sheared. The fiber edges will fuzz, requiring secondary edge finishing to clean them up.

Tools should be hard, clean and sharp. Acetone or an equivalent is used for cleaning the tools.

Cutting and machining composites containing aramid can create a fine fibrous dust (linting) which could be inhaled. A good exhaust system is important.

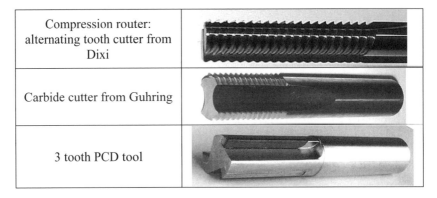

Figure 2.54. *Compression and PCD routers for KFRP part milling*

Composites with resin-to-fiber ratios of 40% or above are usually easier to machine.

2.5.3.2. Routing conditions

The usual routing conditions are given in Table 2.7. Carbide and PCD routers need higher cutting speeds than for the milling of GFRP or CFRP composites. Whatever the router, the feed rate is usually around 0.05 mm/tooth/rev.

A higher cutting speed favors thermal damage while a higher feed rate leads to mechanical damage (delamination, linting).

	Cutting speed Vc, m/min	Feed rate f mm/tooth/rev
Carbide cutters	500-700	0.05
PCD tool	600-800	0.05

Table 2.7. *Cutting conditions for KFRP parts routing*

2.5.4. *Tool wear*

Knowledge of the tool damage mechanisms is a very important aspect of the composite machining as tool wear contributes for a large part to the machining operation costs. The change in tool geometry, a corollary of wear, changes the cutting conditions and therefore damages the machining quality. In the case of FRP machining, tool wear has a mechanical origin (abrasion) and each damage mechanism acts on the tool at different levels depending on the specific machining conditions.

The predominance of one mechanism or the combined effect of several of them depends on the type of machining operation and cutting conditions, and the physicochemical properties of the materials used. The tribological phenomena at the interfaces then control the nature and severity of the wear.

2.5.4.1. *Carbide and diamond-coated carbide tool wear behavior*

During the machining of a GFRP or CFRP, the geometry and the physical state of the cutting are modified by the combined action of the cutting loads and temperatures reached near the cutting edge. The abrasive wear of the carbide tool is then characterized by the formation of striated bands in the contact direction with the workpiece-machined surface or in the direction where the debris slides along the cutting face. The abrasive wear appears on both the rake face (very little, due to the size of the debris) and on the clearance face of the tools (Figure 2.55). This leads to the cutting edge moving back and a loss of sharpness of this edge (the cutting edge rounds). The carbon or glass fiber hard particles are responsible for this abrasive wear.

In the case of diamond-coated carbide tools, the wear mechanisms depend on the adhesion of the coating to the substrate. For a weak bonding, the coating generally does not resist the shear stresses generated at the interfaces between the tool and the material leading to a spalling of the coating from the surface of the tool (Figure 2.56). Then the tool performs as a carbide tool leading to rapid wear by abrasion. There is no significant increase of the tool lifetime in this case.

Figure 2.55. *K20 carbide grade tool after a machined length of 300 m showing significant abrasive wear and low edge acuity [ILI 08]*

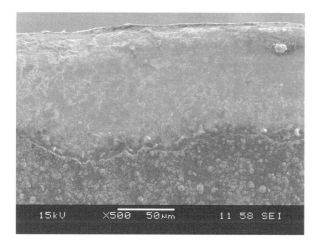

Figure 2.56. *Diamond-coated K20 carbide after a machined length of 50 m highlighting spalling of the coating due to poor adhesion between diamond and carbide [ILI 08]*

In the case of a strong bonding, the diamond coating resists the shear stresses generated at the interfaces leading to an abrasive wear of the coating and increasing the tool lifetime (Figures 2.57 and 2.58).

Figure 2.57. *Balzers® diamond-coated K20 carbide after a machined length of 300 m pointing out very low wear of the coating [ILI 08]*

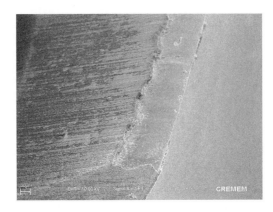

Figure 2.58. *Balzers® diamond-coated K20 carbide after a machined length of 3,200 m highlighting abrasive wear of the coating [ILI 08]*

2.5.4.2. *PCD tool wear behavior*

The wear behavior of PCD tools is highly dependent on 3 parameters:

– the diamond particle size;

– the cobalt content of the binder;

– the diamond type (natural or synthetic) and its level of impurity.

Usually, the higher the diamond particle size and the lower the cobalt content, the longer the tool lifetime. However, the smaller the diamond particle size, the sharper the cutting edge. As a result, the microstructures of the bimodal type of PCD (4 and 25 µm) are preferable to conventional microstructures.

The PCD wear mechanisms most frequently encountered are:

– cleavage of diamond particles of the cutting edge whose graphitic plans are oriented in the direction of the cutting (Figure 2.60);

– cleavage of diamond particles due to dilation of impurities in the grain and a result of the tool heating by the heat generated by cutting (Figure 2.60);

– abrasion of diamond particles and cobalt binder by the hard glass or carbon fibers (Figure 2.59).

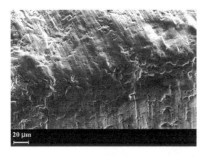
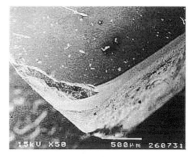

Figure 2.59. *Abrasive wear and attritional edge crumbling [MOS 09]*

Figure 2.60. *Worn-out PCD tool edge [SRE 99]*

2.5.4.3. *Diamond abrasive tool wear behavior*

The wear behavior of diamond abrasive tools is highly dependent on 4 parameters:

– the diamond particle size;

– the diamond particle embedment in the binder;

– the diamond particle volume fraction;

– the diamond type (natural or synthetic) and its level of impurity.

Usually, the higher the diamond particle size and the lower the embedment, the longer the tool lifetime. However, the smaller the embedment, the easier the particle pull out. The higher the particle volume fraction, the longer the tool lifetime.

The diamond abrasive tool wear mechanisms most frequently encountered are:

– cleavage of diamond particles whose graphitic plans are oriented in the direction of cutting (Figure 2.61);

– cleavage of diamond particles due to dilation of impurities in the grain and a result of the tool heating up because of the heat generated by cutting;

– abrasion of diamond particles and nickel binder by the hard glass or carbon fibers (Figures 2.62 and 2.64) with the possibility of inhomogenous wear due to different orientation of fiber plies or layers (Figure 2.63).

Figure 2.61. *Cleavage of a diamond particle*

Figure 2.62. *Regular abrasive wear of part of a diamond particle*

Figure 2.63. *Wave type of abrasive wear*

Figure 2.64. *Abrasive wear of the nickel binder*

2.5.4.4. *Predictive tool wear model*

Wear is often defined as the amount of matter lost by the tool. When wear is characterized by the appearance of ribbed bands formed by abrasion on the clearance face, the criterion for tool life can be established from direct observations on the tool. The friction of the workpiece against the clearance face shows a frontal area of wear, whose height VB is more or less regular. We can quantify the lifetime of tools by a simple measurement of the average width of wear VB.

It is also possible to assess the damage of a cutting tool from indirect criteria based on the performance or quality of machining. Thus, the surface condition and geometric tolerances of the parts can be used as indicators of the level of wear. In drilling, for example, we can define the life of a drill by the number of holes drilled meeting certain quality criteria.

From the wear criteria the lifetime models are established. The oldest and most used is the Taylor model (equation [2.5]) or the modified Taylor model (equation [2.6]):

$$V_c \cdot L^n = C_1 \quad [2.5]$$

$$V_c^p \cdot f^q \cdot w^r \cdot L = C_2 \quad [2.6]$$

These equations describe the relationship between lifetime L and the cutting parameters such as cutting speed Vc, feed rate f and cutting depth w, where C_1, C_2, n, p, q, r are constants to be identified experimentally for each tool/material pair and for each machining process.

In the current literature very few authors have introduced wear into their models.

The model by Tsao *et al.* [TSA 07] takes into account the tool wear but does not include important machining parameters such as the feed rate and cutting speed. It correlates the feed load to the tool wear and the damage of the material. Therefore, a machining quality criterion is used (maximum load allowed to ensure a certain quality).

Lin and Ting [LIN 95] proposed a model for monitoring wear. Feed load F and torque M are fitted as according to the cutting parameters (cutting speed V_c [rpm], feed rate f [mm/rev] or feed speed f_r [mm/min], tool diameter d [mm] and wear w [mm]). The authors developed two model forms to characterize F and M (equations [2.7] to [2.10]):

– Model I:

$$F = a_0 + a_1 \cdot d \cdot f + a_2 \cdot d \cdot w + a_3 \cdot d + a_4 \cdot d^2 \quad [2.7]$$

$$M = b_0 + b_1 \cdot d^2 \cdot f + b_2 \cdot d^2 \cdot w + b_3 \cdot d^2 \quad [2.8]$$

– Model II:

$$\ln(F) = a_0 + a_1 \cdot \ln(V_c) + a_2 \cdot \ln(f_r) + a_3 \cdot \ln(d) + a_4 \cdot \ln(w) \quad [2.9]$$

$$\ln(M) = b_0 + b_1 \cdot \ln(V_c) + b_2 \cdot \ln(f_r) + b_3 \cdot \ln(d) + b_4 \cdot \ln(w) \quad [2.10]$$

The authors noted that the influence of wear on feed load F is more significant than the influence of the wear on torque M. In other words, the recorded signal of the feed load is more sensitive to changes in the tool wear than the torque signal.

Iliescu *et al.* [ILI 08] have developed an interesting model based on the evolution of the feed load with tool wear and composite damage. The model takes into account the evolution of the feed load with feed rate f or feed speed V_f, axial cutting depth p, tool diameter d, cutting speed Vc and tool wear W.

Moreover, the wear is also a function of feed load F_a and contact length Lc (equations [2.11] to [2.13]). Contact length Lc is the length of the path traveled by the tool tip when in contact with the composite material, thus different from the machined length. During a machining sequence i, we must take into consideration all the wear mechanisms related to the machining sequences from $i=1$ to $i-1$.

$$F_{ai} = K \cdot V_{fi}^{a1} \cdot V_{ci}^{a2} \cdot d^{a3} \cdot p_i^{a4} \cdot W_{i-1}^{a5} \qquad [2.11]$$

$$W_i = W_0 + A_0 \cdot \gamma_i \qquad [2.12]$$

$$\gamma_i = \sum_{j=1}^{i} F_{aj} \cdot L_{cj} \qquad [2.13]$$

where W_0 is the initial value of cutting edge acuity, A_0 is the tool abrasion rate, W_i is the tool wear after i machining sequences. Feed load F_i for the ith machining sequence depends on the value of the tool wear reached after i-1 machining trials. K_c is a constant depending on:

− the geometry of the tool;

− the properties of the material being machined.

In the case of CFRP milled with SGS carbide multitooth tools (routing operation), the following tests have been performed in order to determine the parameters of the previous law:

− 1 trial of tests at variable cutting speeds (50, 100 and 200 m/min) and constant feed rate (0.05 mm/rev);

− 3 trials of test at variable feed rate (0.05 , 0.1 and 0.15 mm/rev) and constant cutting speed (100 m/min);

− 1 trial of tests at variable cutting speed (50, 100 and 200 m/min) and constant feed rate (0.05 mm/rev).

Figure 2.65 illustrates the correlation existing between experimental results and the values of the model for these consecutive trials for three different tool diameters. The model is defined in equation [2.14].

$$F_{ai} = 210.46 \cdot V_{fi}^{0.514} \cdot V_{ci}^{-0.653} \cdot d^{0.327} \cdot p_i^{0.803} \cdot \left[2.12 + 0.011 \cdot \gamma_{i-1}\right]^{0.214} \quad [2.14]$$

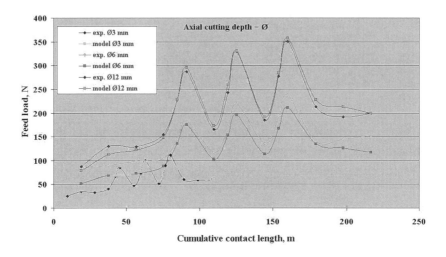

Figure 2.65. *Evolution of the feed load as a function of the contact length Lc for different tool diameters – comparison of experimental data with the wear/feed load model*

The validation of the model allows us to define the tool life behavior as illustrated in Figure 2.66. These results, validated by experiment, point out that the 3 mm diameter tool will break before complete wear while the 12 mm diameter tool resists rupture but leads to burning of the matrix when the feed load begins to be too high.

These types of diagrams allow us to optimize the tool diameter and cutting conditions in order to improve the tool life while respecting the integrity of the material.

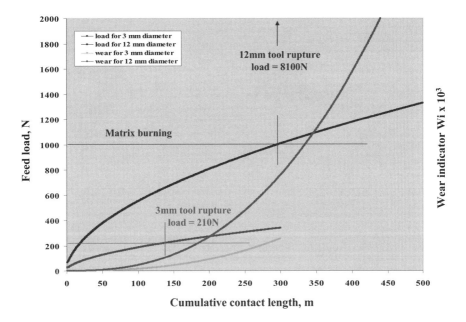

Figure 2.66. *Wear indicator Wi and feed load according to the contact length Lc and tool diameter (cutting speed Vc= 200 m/min; feed rate f = 0.05 mm/rev; axial cutting depth = tool diameter)*

2.6. Turning of composite materials

2.6.1. *Tool geometries*

The turning remains a marginal operation in the machining of FRP materials compared to the processes of drilling or trimming. However, it is a process necessary to achieve the functional surfaces of revolving parts. As for milling, the tools used and that give the best results are essentially carbide and diamond-coated carbide inserts and the PCD inserts (Figures 2.67 and 2.68). The considerations concerning the sharpness of the cutting edge during the milling of GFRP, CFRP and KFRP materials remain valid for turning operations. Table 2.8 provides the essential geometric characteristics of tools for turning of GFRP, CFRP and KFRP materials.

102 Machining Composite Materials

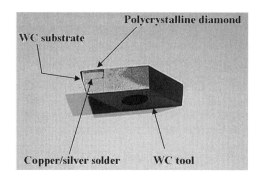

Figure 2.67. *Constitution of a PCD insert*

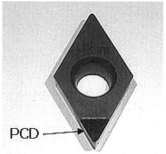

Figure 2.68. *PCD insert for turning*

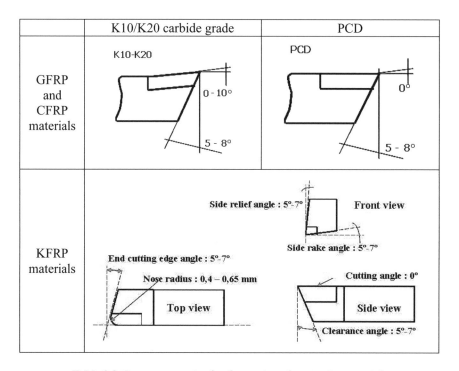

Table 2.8. *Insert geometries for the turning of composite materials*

2.6.2. Turning conditions

The usual turning conditions are given in Table 2.9 for GFRP and CFRP materials and in Table 2.10 for KFRP materials. Recommended cutting speeds are in the range between 50 and 250 m/min for carbide inserts and between 200 and 800 m/min for PCD inserts turning GFRP or CFRP materials.

As opposed to milling, the feed rates can be increased up to 0.3 mm/rev depending on the composite texture and the turning sequence. If the tool should leave the workpiece, this feed rate is high enough to produce the delamination or edge effects.

	Cutting speed Vc, m/min	Feed rate f mm/rev	Cutting depth mm
K10/K20 carbide grade	50-250	0.05-0.3	< 8
PCD	200-800	0.05-0.3	< 7

Table 2.9. *Turning conditions for GFRP and CFRP composites*

	Cutting speed Vc, m/min	Feed rate f mm/rev	Cutting depth mm
K10/K20 carbide grade	75-150	0.04-0.06	0.25-0.50
PCD	400-800	0.04-0.06	0.25-1

Table 2.10. *Turning conditions for KFRP composites*

As in milling, PCD inserts behave better than carbide or diamond-coated carbide inserts. During the turning of GFRP materials, the tool life of the PCD insert is 800 times longer than the tool life of the carbide insert (for a cutting speed of 200 m/min) and 40 times longer than the tool life of the diamond-coated insert (Figure 2.69).

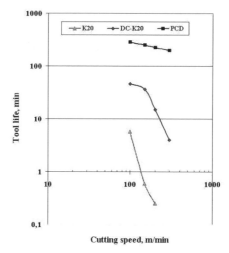

Figure 2.69. *Tool life for carbide, diamond-coated carbide and PCD tools according to cutting speed in turning of GFRP (Vf = 35%)*

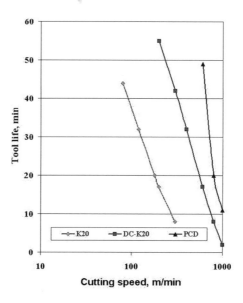

Figure 2.70. *Tool life for carbide, diamond-coated carbide and PCD tools according to cutting speed in turning of CFRP (Vf = 40%)*

In the case of CFRP materials, the gains are lower mainly due to the carbon fiber properties. However, for the same tool life (30 min), the increase in productivity (cutting speed) is around 9 between the carbide insert and the PCD insert (100 to 900 m/min; see Figure 2.70).

2.6.3. *Surface quality*

As for milling operations, the surface quality of the part after turning depends on:

– the orientation of the fibers with respect to the cutting direction. The cutting mechanisms (point out in section 2.3.1) appears depending of the texture of the composite: fiber buckling when oriented at 0° (Figure 2.71), fiber shearing when oriented at +45° (Figure 2.72) or fiber pull out when oriented at -45° (Figure 2.73);

– the turning conditions;

– the sharpness of the cutting edge.

When all these parameters are under control and well chosen, a machined surface without any damage can be achieved (Figure 2.74).

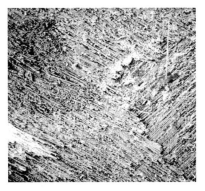

Figure 2.71. *Fiber buckling for orientation at 0°*

Figure 2.72. *Fiber shearing for orientation at +45°*

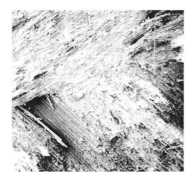 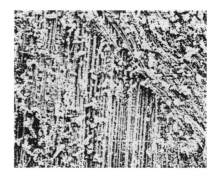

Figure 2.73. *Fiber pull out for orientation at -45°*

Figure 2.74. *Undamaged surface after a turning sequence with PCD tool (Vc= 800 m/min; f= 0.05 mm/rev; rε= 2 μm)*

2.7. Conclusions

In this chapter the main aspects of the machinability of polymer matrix composites have been presented. The key factors to be taken into account are that:

– although the PCD tool price is higher than that of the carbide or diamond-coated carbide tools, the performances of such tools are higher if the cutting conditions are well chosen;

– diamond-coated carbide tools are a cheaper alternative to PCD tools. The improvement in adhesion of the diamond coating on the carbide substrate makes it possible to increase the performances of these tools (tool lifetime) gradually;

– the edge acuity or sharpness is a key factor for tool lifetime. Edge finishing techniques such as laser preparation will in the future enable us to improve the robustness of the cutting edge and to increase lifetime.

At the same time, the optimization of the multi-edge routing tools has been presented, taking into account the tool material, geometry, edge shape, coating, and cutting conditions. After the testing of coated and uncoated milling tools several conclusions can be pointed out:

– Multitooth milling tools made with submicron grade carbide with substrate 6% Co, coated with a monolayer 4 µm-thick of TiAlN, results in the best tool life, reaching 40-50 m before bad cutting is detected. The wear is basically abrasion, therefore the thicker the coating layer, the longer the tool life. Around 4 µm is a recommended value. More than this thickness would affect the sharpness of cutting edges.

– Carbon+aramid fiber-reinforced composites are more difficult to machine than those reinforced only with carbon fibers. The specific cutting force can be an indicator of the machinability of each composite; however, other parameters like the type of fibers are also important.

2.8. Acknowledgments

Thanks are due to the MECO and TRI_COM projects supported by the Basque government (INNOTEK), and ETORTEK 2008 also from the Basque government. Thanks also to the firms DYFA (end user), Metal Estalki (Platit coatings), Kendu (milling tools) and Aernnova (composites and aeronautical parts) for the technical suggestions and support, and special thanks to Raul Fidalgo and Asier Fernández for their help in testing.

Part of this work has also been supported by the French Department of Education and Science through the MEDOC project (contract n°02K0538) and the Common Fund for Aquitaine/Euskadi Cooperation.

2.9. References

[ABR 92] S. ABRATE, D.A. WALTON, "Machining of composite materials: part I: traditional methods", *Composites Manufacturing*, Vol. 3, pp. 75-83, 1992.

[BHA 95] N. BHATNAGAR, N. RAMAKRISHNAN, N.K. NAIK, R. KOMANDURI, "On the machining of fiber reinforced plastic (FRP) composite laminates", *International Journal of Machine Tools and Manufacture*, Vol. 35, no. 5, pp. 701-716, May 1995.

[BHA 98] D. BHATTACHARYA, D.P.W. ROIGAN, "A study of hole drilling in kevlar composites", *Composites Science and Tech.*, Vol. 58, pp 267-283, 1998.

[CAP 08] E. CAPELLO, A. LANGELLA, L. NELE, A. PAOLETTI, L. SANTO, V. TAGLIAFERRI, "Drilling polymeric matrix composites", *Machining, Fundamentals and Recent Advances*, pp. 167-194, Springer Verlag, 2008.

[CHA 95] A. CHAMBERS, G. BISHOP, "The drilling of carbon fiber polymer matrix composites", *Proceedings of ICCM-10*, Whistler, B.C., Canada, 1995.

[DAV 04] J.P. DAVIM, P. REIS, C.C. ANTÓNIO, "Drilling fiber reinforced plastics (FRPS) manufactured by Hand Lay-Up: influence of matrix (Viapal VUP 9731 and ATLAC 382-05)", *J. Material Processing Techn.*, 155–156, pp. 1828–1833, 2004.

[DAV 05] J.P. DAVIM, P. REIS, "Damage and dimensional precision on milling carbon fiber-reinforced plastics using design experiments", *J. of Materials Processing Techn.*, Vol. 160, pp. 160-167, 2005.

[GHI 03] P. GHIDOSSI, Contribution à l'étude de l'effet des conditions d'usinage d'éprouvettes en composites à matrice polymère sur leur réponse mécanique, PhD N°2003-19, Procédés de Fabrication, ENSAM Chalons en Champagne, 2003.

[GIR 97] F. GIROT, "High-speed abrasive milling of ceramic matrix composite materials", *1ère Conférence franco-allemande sur l'usinage grande vitesse*, A. MOLINARI (ed.), Metz, France, pp. 351-356, June 1997.

[GUE 94] P. GUÉGAN, Contribution à la qualification de l'usinage de matériaux composites à matrice organique, PhD thesis no. 2025, Mechanical engineering, E.C. Nantes, 1994.

[HIC 87] J. HICKEY, "Drilling graphite composites", *Modern Machine Shop*, Gardner Publication, Cincinnati, Ohio, pp. 84-90, March 1987.

[HO-C 90] H. HO-CHENG, C.K.H. DHARAN, "Delamination during drilling in composite laminates", *Trans. of the ASME, Journal of Engineering for Industry*, Vol. 112, pp 236-239, 1990.

[HO-C 92] H. HO-CHENG, H.Y. PUW, K.C. YAO, "Experimental aspects of drilling of some fiber-reinforced plastics", *Proceedings of the Machining of Composite Materials Symposium*, ASM material week, Chicago, Illinois, pp. 127-138, 1-5 November, 1992.

[ILI 08] D. ILIESCU, Approches expérimentales et numériques de l'usinage à sec des composites carbone/époxy, PhD N°2008 ENAM 0045, Sciences des Métiers de l'Ingénieur, Arts et Métiers ParisTech, Bordeaux, 2008.

[INO 97] H. INOUE, E. AOYAMA, T. HIROGAKI, K. OGAWA, H. MATSUSHITA, Y. KITAHARA, T. KATAYAMA, "Influence of tool wear on internal damage in small diameter drilling in GFRP", *Composite Structures*, Vol. 39, pp. 55-62, 1997.

[JAN 03] J. JANARDHAR, Tool wear of diamond interlocked tools in routing of composites, PhD Thesis, Wichita State, 2000.

[KEN 08] B. KENNEDY, "Competently cutting composites", *Cutting Tool Engineering*, Vol. 60, Issue 7, 2008.

[KÖN 84] W. KÖNIG, P. GRASS, A. HEINTZE, "Comment optimiser l'usinage du Kevlar", *Machine Moderne*, Société de publications mécaniques, Paris, Vol. 891, p. 8, 1984.

[KÖN 89] W. KÖNIG, P. GRASS, "Quality definition and assessment in drilling of fiber reinforced thermosets", *Annals of the C.I.R.P.*, Vol. 38, pp. 119-124, 1989.

[KOP 80] A. KOPLEV, "Cutting of CFRP with single edge tools", *Third International Conference on Composite Materials*, Paris, pp. 1597 - 1605, 1980.

[KOP 83] A. KOPLEV, A. LYSTRUP, T. VORM, "The cutting process, chips and cutting forces in machining CFRP", *Composites*, Vol. 14, no. 4, pp. 371-376, 1983.

[KRI 92] R. KRISHNAMURTHY, G. SANTHANAKRISHNAN, S.K. MALHORTA, "Machining of polymeric composites", *Proceedings of the Machining of Composite Materials Symposium*, ASM material week, Chicago, Illinois, pp. 139-148, 1-5 November, 1992.

[LAC 01] F. LACHAUD, R .PIQUET, F. COLLOMBET, L. SURCIN, "Drilling of composite structures", *Composite Structures*, Vol. 52, 511-516, 2001.

[LIN 95] S.C. LIN, C.J. TING, "Tool wear monitoring in drilling using force signals", *Wear*, Vol. 180, no. 1-2, pp. 53-60, 1995.

[LOP 09] L.N. LÓPEZ DE LACALLE, A. LAMIKIZ, F. CAMPA, A. FERNÁNDEZ, I. ETXEBERRIA, "Design and test of a multitooth tool for CFRP milling", *Journal of Composite Materials*, forthcoming, 2009.

[MOS 09] S.G. MOSELEY, K.-P. BOHN, M. GOEDICKEMEIER, "Core drilling in reinforced concrete using polycrystalline diamond (PCD) cutters: wear and fracture mechanisms", *Int. Journal of Refractory Metals & Hard Materials*, Vol. 27, pp. 394-402, 2009

[PIQ 99] R. PIQUET, Contribution à l'étude des réparations provisoires structurales aéronautiques; étude du perçage de plaques minces en carbone/époxy, PhD Thesis, no. 3339, UPS Toulouse III, 1999.

[RAH 99] M. RAHMAN, S. RAMAKRISHNA, J.R.S. PRAKASH, D.C.G. TAN, "Machinability study of carbon fiber reinforced composite", *Journal of Materials Processing Technology*, Vol. 89-90, pp. 292-297, 1999.

[RAM 94] M. RAMULU, E. ROGERS, "Simulation of router action on a lathe to test the cutting tool performance in edge-trimming of graphite/epoxy composite", *Experimental Techniques*, Vol. 18, pp. 23-35, 1994.

[RAM 99] M. RAMULU, "Cutting-edge wear of polycrystalline diamond inserts in machining of fibrous composite material, Machining of Ceramics and Composites", *Manufacturing Engineering and Material Processing*, Dekker, 1999.

[RAO 08] G.V.G. RAO, P. MAHAJAN, N. BHATNAGAR, "Orthogonal machining of UD-CFRP composites", *International Journal of Materials and Product Technology*, Vol. 32, pp. 168-187, 2008.

[SCH 97] H. SCHULZ, "Fraisage grande vitesse des matériaux métalliques et non métalliques", *Sofetec*, Chap. 4.7, pp. 175-201, 1997.

[SEI 07] M.A. SEIF, U.A. KHASHABA, R. ROJAS-OVIEDO, "Measuring delamination in carbon/epoxy composites using a shadow moiré laser based imaging technique", *Composite Structures*, Vol. 79, pp. 113-118, 2007.

[SRE 99] P.S. SREEJITH, R. KRISHNAMURTHY, K. NARAYANASAMY, S.K. MALHOTRA, "Studies on the machining of carbon: phenolic ablative composites", *Journal of Materials Processing Technology*, Vol. 88, pp. 43-50, 1999.

[TSA 07] C.C. TSAO, H. HOCHENG, "Effect of tool wear on delamination in drilling composite materials", *International Journal of Mechanical Sciences*, Vol. 49, no. 8, pp. 983-988, 2007.

[WAN 95] D.H. WANG, M. RAMULU, D. AROLA, "Orthogonal cutting mechanisms of graphite/epoxy composite. Part I and Part II", *International Journal of Machine Tools and Manufacture*, Vol. 35, Issue 12, pp. 1623-1648, December 1995.

[WEN 97] C. WEN-CHOU, "Some experimental investigations in the drilling of carbon fiber-reinforced plastic (CFRP) composite laminates", *Int. J. of Machine Tools and Manufac.*, Vol. 37, pp. 1097-1108, 1997.

Chapter 3

Drilling Technology

This chapter is focused on the principal aspects concerning the drilling of reinforced polymeric matrix composites. Initially, the particular behavior of this grade of material when subjected to drilling – due to its unique properties – is discussed, followed by a comment on the performance of distinct tool materials and geometries. Next, the principal wear mechanisms involved when drilling polymeric composites are reviewed and the drilling forces and torque are depicted. Finally, the influence of the above-mentioned factors on the quality of the machined component, assessed in terms of damage, surface texture and dimensional and geometric deviations are detailed.

3.1. Introduction

Drilling is probably the machining operation most widely applied to polymeric matrix composites owing to the need to assemble components, produced mainly as laminates, through riveting and bolting. For example, 100,000 holes are required in a small engine aircraft and millions of holes are necessary for a large transportation aircraft [ELS 04]. Consequently, tool material and geometry as well as

---

Chapter written by Alexandre M. ABRÃO, Juan C. CAMPOS RUBIO, Paulo E. FARIA and J. Paulo DAVIM.

the machining parameters can significantly affect the costs of the operation and the performance of the machined component.

Epoxy and polyester are the principal thermosetting matrix materials whereas glass, carbon and to a lesser extent aramid fibers are the main reinforcing phases used in polymeric composites. Initially developed for aerospace applications owing to their thermal properties [WIN 97], the use of carbon-carbon composites has increased dramatically for high performance applications.

The combination of two insoluble (matrix and reinforcing) materials results in outstanding properties (namely high specific strength and specific modulus) that are useful for the manufacture of products as distinct as printed wiring boards and rocket nozzles. Adams and Mahery [ADA 03] point out the potential of polymeric composites for vibration damping and discuss the influence of fiber orientation and stacking on the specific damping capacity of composites.

Nevertheless, this combination results in an anisotropic and non-homogenous material, which represents a challenge for the production of high quality goods at an acceptable cost. Furthermore, the orientation, volume fraction and form (continuous or discontinuous fibers) of the reinforcing phase can drastically affect machining. In the case of short fibers, some properties such as strength, modulus and toughness increase with fiber length [FU 00]. According to [GOR 03], the higher the fiber volume fraction, the higher the modulus of elasticity, strength and brittleness of the composite. On the other hand, the higher the matrix fraction, the higher the toughness and thermal expansion coefficient and the lower the density, strength, stiffness and thermal stability. Gliesche *et al.* [GLI 05] report that the thinner the reinforcing layer, the higher the values of the mechanical properties.

The influence of the fiber volume fraction on the mechanical properties of a polyamide matrix reinforced with unidirectional carbon fibers was investigated by [BOT 03]. The findings indicated that the longitudinal and transverse tensile modulus and the compression strength increased with the fiber volume fraction, but the results of the interlaminar shear strength and compressive shear tests were inconclusive. In addition to that, composites with lower matrix volume fraction presented fewer voids and applying loads corresponding to 20% and 30%, respectively, of the tensile and compressive rupture loads resulted in internal delamination of the composite samples.

The effect of the shape of the cross-section of carbon fibers (used as reinforcing in an epoxy matrix) on the properties of the composite was studied by [PAR 03]. In addition to solid round fibers, hollow round and C-shaped fibers were tested. The findings indicated that the C-shaped fibers presented superior performance, probably owing to the fact that the increased contact surface with the matrix improved interfacial binding.

Furthermore, the higher contact area of C-shaped fibers resulted in a higher load transfer capacity. As a result, similar values of tensile strength were obtained for the three materials, but the tensile modulus of the composite reinforced with hollow round fibers was highest and the torsional rigidity of the C-shaped and hollow round composites were similar and presented a twofold increase in comparison with the composite reinforced with solid round fibers. Finally, the composite reinforced with C-shaped carbon fibers presented superior damping factor, interlaminar shear strength and transverse flexural strength.

Bhatnagar *et al.* [BHA 95] used the Iosipescu shear test in order to investigate the influence of fiber orientation on the shear strength of unidirectional carbon fiber-reinforced epoxy composite. The results indicated that the shear strength decreases as the fiber angle increases. The authors state that this method can be used to predict the shear force for orthogonal cutting with an acceptable agreement, as long as the fiber orientation angle is smaller than 60°.

The mechanisms associated with chip formation when cutting unidirectional polymeric reinforced composite were studied by [WAN 95], who found that the cutting mechanisms and chip formation are closely dependent on fiber orientation and tool geometry.

As far as the fiber orientation is concerned, three cutting mechanisms were observed: fracture along the fiber/matrix interface due to cantilever bending followed by fracture perpendicular to the fiber direction (for a fiber orientation of 0°); compression induced shear followed by fracture along the fiber/matrix interface (for positive fiber orientation angles below 75°) and shear fracture along the fiber/matrix interface with deformation induced by compressive load (for 90° and negative fiber orientation angles).

Figure 3.1 shows the chip formation mechanism when cutting samples of epoxy resin reinforced with woven glass, carbon and aramid fibers (Figures 3.1a, 3.1c and 3.1d, respectively), subjected to quick stop tests using a planning machine and a diamond-coated carbide insert as a cutting tool. While the glass and carbon fiber-reinforced materials, which are relatively brittle, present a similar pattern showing the region corresponding to the cutting tool wedge fairly and, for the glass fiber-reinforced composite, the crack propagation ahead of the cutting edge, in the case of the aramid-reinforced composite, which is tougher, the fuzzy nature of the fibers does not allow the observation of the chip formation mechanism.

Comparing the photographs shown in Figure 3.1, we may have an indication of the difficulty expected when machining aramid fiber-reinforced polymeric composites. In general, the following mechanisms take place when cutting fiber-reinforced polymeric composites [SAN 88]: plastic deformation, shearing and bending rupture. However, due to the high flexibility of aramid fibers, bending rupture is absent and plastic deformation and fiber elongation predominate.

In the case of glass fiber-reinforced composites, the fibers are more brittle and less flexible than aramid fibers. Consequently, cutting takes place by the three above-mentioned mechanisms. Finally, carbon fibers are the most brittle and possess the least strain at failure; therefore, they become crushed and fracture sharply.

(a) Glass fiber-reinforced epoxy   *(b) Detail of (a)*

(c) Carbon fiber-reinforced epoxy   *(d) Aramid fiber-reinforced epoxy*

Figure 3.1. *Scanning electron microscopy samples of quick stop machining tests for epoxy resin composites reinforced with: (a) glass fibers; (b) glass fibers (detail of a); (c) carbon fibers and (d) aramid fibers*

3.2. Standard and special tools

Both tool geometry and material are critical aspects that must be carefully taken into account when machining composite materials owing to the fact that they significantly affect machining forces and, consequently, the quality of the machined component and the economics of the operation (tool wear and machining parameters). In addition to that, the inappropriate selection of the tool geometry results in higher temperature due to friction, thus promoting high wear rates and cutting forces. According to [CHE 97], the judicious selection of the tool geometry and corresponding machining parameters may result in the production of holes free of defects. In the particular case of the drilling operation, the drill is a complex tool and its cutting edges (lips) present variable rake, inclination and clearance angles. Furthermore, the material at the center of the hole is extruded by the chisel edge (web) owing to the fact that the cutting speed tends towards zero as the distance from the drill center is reduced.

When drilling fiber-reinforced composites, high-speed steel and ISO grade K10 and K20 cemented carbide are the principal tool materials, being employed in the same proportion [ABR 07]. In spite of their superior hardness and wear resistance, diamond-based (mono and polycrystalline) tools and polycrystalline cubic boron nitride (PCBN) are seldom cited in the published literature. Oxide and non-oxide ceramics are not employed in drilling operations due to the limitations related to the manufacture of a tool with geometry as complex as that of a drill.

High-speed steel is widely used for both standard drills and tools especially designed for the machining of composites, but, attention must be paid to the abrasive nature of the reinforcing fibers [TSA 07], which may lead to high wear rates and an increase in the thrust force and in the damage of the machined component. According to [DAV 03a], minimal damage in composite laminates is observed when drilling with carbide drills in comparison with high-speed steel tools. Furthermore, pre-drilling is frequently recommended when drilling composites with standard point twist drills in order to minimize the chisel edge effect, thus improving the quality of the hole [TSA 03].

In contrast to drilling, turning usually allows the use of a wider range of cutting tool grades. The performance of three distinct tool materials (uncoated carbide, ceramic and PCBN) when turning a carbon fiber-reinforced composite was investigated by [RAH 99].

The results indicated that the PCBN compact outperformed the remaining tool grades, while the ceramic tool was unable to withstand the thermal and mechanical shocks. Similar work was conducted by [FER 01], who conducted turning tests on a carbon-carbon composite using the following cutting tool grades: uncoated and coated carbide, whisker-reinforced and mixed alumina, polycrystalline diamond

(PCD) and PCBN. Nevertheless, the findings indicated that the coated carbide tool gave the best results when roughing, whereas the PCD was superior when finish turning. Turning tests on a glass fiber-reinforced epoxy composite using three tool materials (monocrystalline diamond, PCD and PCBN) and two cutting edge preparations (nose radius of 0.5 mm and straight edge of 1.5 mm) indicated that the best results with regard to surface finish and turning forces were obtained using the straight edge preparation [AN 97]. As far as the tool grades are concerned, the best results were obtained with the monocrystalline diamond. Comparable results were given by the PCD and PCBN compacts.

Various attempts have been made in order to identify the optimal tool geometry for drilling fiber-reinforced composites and thus to reduce thrust force, torque and tool wear and to improve the quality of the machined surface. In addition to standard point twist drill and helical point drill, special geometry drills such as step drill, core drill, negative point angle drill (also known as candle stick or multifaceted drill) and parabolic drill have been used together with drills having alterations in their original geometry.

Reports by [PIQ 00] and [BHA 04] say that increasing the number of cutting edges tends to reduce thrust force and torque, thus minimizing the damage in the composite. In contrast, [DAV 03a] compared the performance of a twist drill with a four flute drill and noticed that the former was responsible for less delamination on a carbon fiber-reinforced laminate. Similarly, [LIN 96] reported that a carbide twist drill provided lower thrust force and torque compared with a negative point angle carbide drill when drilling carbon fiber-reinforced epoxy composite.

A reduction of 50% in the thrust force and 10% in torque in comparison to a conventional twist is claimed when a trepanning tool (core drill) is used to drill glass fiber-reinforced composite [MAT 99]. Furthermore, increasing the twist drill diameter results in an appreciable elevation in the thrust force and torque due to the higher shear area, which is not observed using a trepanning tool [ELS 04].

The influence of the drilling-induced damage on the mechanical properties of carbon-reinforced epoxy composite coupons was investigated by [PER 97]. The damage was induced by drilling the specimens using three distinct tools: a diamond grain-coated mandrel, a PCD-tipped drill and a cemented carbide dagger drill.

In general, the findings indicated that the diamond-coated mandrel and the dagger drill provided similar results, i.e. superior strength and fatigue life compared with the PCD-tipped drill.

A comprehensive investigation on the influence of the drill geometry on the machinability of a carbon fiber-reinforced composite was conducted by [HOC 06].

Five distinct drill geometries (twist drill, saw drill, negative point angle drill, core drill and step drill) were tested and the findings indicated that using the core drill higher feed rates could be employed without the risk of delamination. In contrast, the lowest feed rate should be employed for the twist drill in order to avoid damaging the composite.

Figure 3.2 presents four distinct drill geometries typically employed for drilling fiber-reinforced composite materials: two helical point drills with distinct edge preparations (Figures 3.2a, 3.2b and 3.2c), a three flute drill (Figure 3.2d) and a negative point angle drill (Figures 3.2e and 3.2f).

According to Shaw [SHA 84], the geometry of a drill represents a compromise between conflicting requirements, including: reducing the chisel edge, which leads to lower thrust force, but the resistance to chipping and the torsional rigidity are reduced; larger flutes provide larger areas for conveying chips at the expense of lower torsional rigidity; finally, an increase in the helix angle allows the quicker removal of chips, although the strength of the cutting edges is diminished.

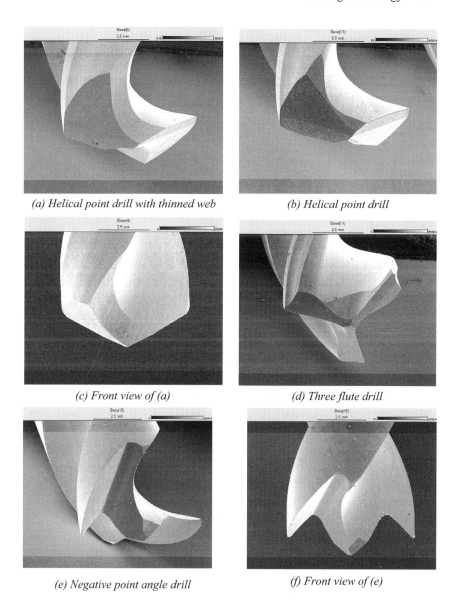

(a) Helical point drill with thinned web *(b) Helical point drill*

(c) Front view of (a) *(d) Three flute drill*

(e) Negative point angle drill *(f) Front view of (e)*

Figure 3.2. *Drill geometries typically employed for drilling fiber-reinforced composite (Ø5 mm)*

Four distinct drill geometries were tested by [SIN 06] against unidirectional glass fiber-reinforced epoxy resin: four and eight facet drills, a parabolic drill and a step drill. The results based on thrust force, torque and damage area indicated that the eight facet drill and the step drill gave similar results, superior to the other geometries tested. The four facet drill was not recommended for machining of unidirectional glass fiber-reinforced laminates.

In addition to the anisotropy and lack of homogenity observed in glass and carbon fiber-reinforced polymeric composites, aramid fibers present higher toughness and flexibility, which makes machining of these composites a more complex task. In order to overcome such drawbacks, Bhattacharyya and Horrigan [BHA 98] conducted a series of drilling experiments at room and cryogenic temperatures and compared the behavior of high-speed steel with standard (twist drill) and modified (negative point angle drill) geometries. Moreover, the drills were tested with two helix angles. The results indicated that, in general, the negative point angle drill outperformed the standard twist drill and the use of liquid nitrogen promoted longer tool lives and a better machined surface finish, albeit with the thrust force and delamination increased. The superior performance of a negative point angle drill when machining glass and carbon fiber-reinforced composites was reported by [DAV 03b] and [DAV 04a], who observed lower delamination and thrust force values compared with helical drills.

The principal features of a drill subjected to alterations in its geometry aiming to improve its performance when cutting fiber-reinforced polymeric composites are the rake, clearance, point and helix angle as well as the chisel length. Experimental work concerned with the influence of the tool point angle (ranging from 75° to 160°) on the quality of carbon fiber-reinforced epoxy composite suggested that using the lowest point angle value resulted in minimal delamination owing to the reduction in thrust force, but nevertheless, the surface roughness values were barely acceptable [ENE 01].

Piquet *et al.* [PIQ 00] compared the performance of a standard twist drill (helix angle of 25°, major cutting edge angle of 59° and clearance and rake angles of 6°) with a special geometry drill (three cutting edges, major cutting edge angle of 59°, helix and rake angles of 0° and clearance angle of 6°) possessing a minor cutting edge angle varying from 59° to 0°. The superior performance of the special geometry drill was attributed to the smaller contact length with the hole wall.

The effect of the point angle, helix angle and chisel edge angle on thrust force and torque when drilling carbon fiber-reinforced composite with high-speed steel drills was investigated by [CHE 97], who concluded that by increasing the point angle the thrust force is elevated, although torque is reduced. On the other hand, when the helix angle and the chisel edge angle are increased, the values of both thrust force and torque are reduced.

3.3. Cutting parameters

The selection of tool grade and geometry greatly affects the choice of cutting speed and feed rate to be employed when drilling fiber-reinforced polymeric laminates. However, in contrast to metals, drills used to machine polymeric composites can simultaneously withstand high cutting speeds and feed rates. Additionally, the choice of drill diameter depends on the design of the component to be machined rather than on the optimal cutting conditions and determines, together with cutting speed and feed rate, the metal removal rate and, consequently, drilling power. Nevertheless, a number of reports such as [TSA 03], [ELS 04] and [OGA 97] point out the importance of pre-drilling in order to reduce damage in composite materials.

Typical values for the cutting speed generally range from 20 to 60 m/min and 0.3 mm/rot is the maximum value for feed rate usually found in the published literature when drilling fiber-reinforced composites with high-speed steel and tungsten carbide drills [ABR 07]. These conservative values may be explained by the following reasons: in the case of cutting speed, higher values may cause the softening of the matrix and clogging of this phase on the cutting tool. On the other hand, increasing cutting speed resulted in lower thrust force and torque values owing to the higher temperature associated with the low coefficient of thermal conduction and the low glass transition temperature of plastics [KHA 04]. In addition to that, owing to the fact that the diameter of the drills rarely exceeds 10 mm, the use of higher cutting speeds may require rotational speeds which are not available in most machine tools. As far as the feed rate is concerned, this is the principal parameter affecting the damage induced in the composite; therefore, lower values are imperative for the production of components with acceptable quality. According to [TSA 07], the use of high-speed steels as cutting tool material results in high wear rates, which promotes higher thrust force and, consequently, severe damage in the composite. Therefore, low feed rates should be employed, especially as tool wear progresses.

The influence of drill diameter on thrust force and torque when drilling glass fiber-reinforced polymeric composites with standard high-speed steel was studied by [ELS 04]. An increase in drill diameter and feed rate resulted in an elevation of thrust force and torque due to the increase in the shear plane area. In contrast to work by [KHA 04], an increase in cutting speed resulted in higher thrust force and torque, but not to the same extent as when feed rate was elevated. The reason for that may be related to the accelerated wear rates observed when high-speed steel drills are employed.

However, when turning, cutting speed increases with the diameter of the work material and ultra-hard tool materials (PCD and PCBN) are more easily available in the required format and geometry. Machining of a carbon-reinforced composite can thus be carried out with coated carbide and PCD inserts at critical cutting speeds of 100 and 300 m/min, respectively, without damaging the component [SRE 99].

Several attempts have been made to increase production rates without impairing the quality of the machined part, of which two are noteworthy: the use of a back support and variable feed rate devices. The long-standing use of backing has been reported by [INO 97], [BHA 98] and [CAP 04], as it avoids the fracture of the last layers of the laminate by the drill point, thus preventing delamination as the drill exits the laminate. The principal drawback associated with the use of a back support is related to the fact that the back material is bonded to the laminate, thus increasing production time.

The use of devices that allow the progressive reduction of feed rate as the drill point approaches the exit is more ingenious and may be applied to conventional drilling machines [CAP 04]. The use of vibratory drilling at low frequency and high amplitude has also been claimed as an alternative to avoid damage in the workpiece [ARU 06]. Nevertheless, the widespread use of numerical control machines has made the feed speed control simpler and eliminated additional devices [DHA 00]. Alternatively, an artificial neural network can be employed in order to determine the required feed speed based on the least hole damage or thrust force; see [STO 96] and [TSA 08]. In all the previously mentioned cases, production rate is not as drastically affected as by the use of a back support.

3.4. Tool wear

In contrast to the cutting of metals, where it possesses a secondary role in tool wear compared with diffusion and adhesion, abrasion is considered a major wear mechanism when machining fiber-reinforced polymeric composites.

Figure 3.3 shows grooves on the major and minor clearance faces caused by friction with the fibers present in the reinforcing phase. Furthermore, the sharpness of both major and minor cutting edges has been impaired, as suggested by the edge rounding. Tool wear appears to progress at approximately the same rate at the rake and clearance faces, higher wear rates being observed when tools with a lower clearance angle are used [CAP 96].

According to [SRE 99], abrasion and fatigue are the principal wear mechanisms observed when machining carbon fiber-reinforced composites, the latter being due to the difference between the strength values of the matrix and reinforcing materials. In addition to that, the authors report that there is a critical cutting speed above which tool wear rate increases drastically owing to the elevation of temperature. As a consequence of the deterioration of the cutting edge, machining forces also tend to increase.

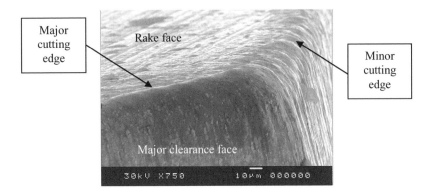

Figure 3.3. *Abrasive wear observed on high-speed steel after drilling fiber-reinforced polymeric composite*

The wear mechanisms involved when face turning glass, carbon and aramid fiber-reinforced epoxy composite with tungsten carbide inserts were investigated by [SAN 88]. Severe nose wear and edge rounding were observed after machining the glass fiber-reinforced material. Abrasion and cobalt depletion due to high temperatures are considered the principal wear mechanisms. In the case of carbon fiber-reinforced composite, uniform nose, flank and crater wear were observed and carbide particles once present in the tool were removed due to their affinity with the carbon fibers. Finally, machining of aramid fiber-reinforced composite resulted in chip notching due to the continuous rubbing of the loosened fibers against the major cutting edge.

The wear mechanisms present when face turning glass fiber-reinforced epoxy composite with high-speed steel is characterized by abrasion together with either chipping of tools which possess low fracture rupture strength, or plastic deformation of the cutting edge for tools with higher fracture rupture strength [SAN 89]. Moreover, the cobalt content significantly affects tool wear.

Rahman *et al.* [RAH 99] observed excessive notch wear on cemented carbide tools when turning carbon fiber-reinforced composites. In addition to that, they noticed that ceramic tools were not recommended for machining these materials due to the fact that they are prone to chipping caused by the thermal and mechanical shocks.

Useful information concerning the tribological behavior of the materials under controlled conditions can be obtained through ball-on-disc testing. These data can be used to predict the behavior of the cutting tool/work material pair during machining. The ball-on-disc test provides an accurate estimate of adhesive wear, which is a wear mechanism expected to be observed during the machining of composite materials, together with abrasive wear. Ball-on-disc tests using high-speed steel and tungsten carbide counter balls against glass fiber-reinforced epoxy composite discs were conducted by [FAR 08]. The findings indicated that, for a sliding distance of 250 m, the friction coefficient the high-speed steel increased drastically, ranging from 0.13 to 0.56, while for the tungsten carbide this parameter remained steady with an average value of 0.15. Similarly, the wear depth for the high-speed steel increased steeply, suggesting poor wear resistance against the composite. In contrast, the wear depth observed using the tungsten carbide counter ball did not change during the test. According to the authors, these results indicated that the adhesive wear and plastic deformation that took place on the high-speed steel and composite surfaces produced wear debris and transferred particles, thus altering the geometry of the contact surfaces.

The wear developed on high-speed steel, cemented carbide and titanium nitride coated carbide drills (diameter of 5 mm) after drilling 1,000 holes in glass fiber-reinforced epoxy laminates at a cutting speed of 86 m/min and feed rate of 0.15 mm/rot can be seen in Figure 3.4. While the high-speed steel drill presents an appreciable loss of material on the major cutting edge (indicated by the rounding of the cutting edge), the uncoated and coated carbide drills show a similar wear pattern, characterized by shallow grooves on the major clearance faces. In contrast to the work by [FER 01] on the turning of a carbon-carbon composite, the presence of a coating did not represent a substantial improvement in the wear resistance of the tool. The reason for that may be the fact that the temperatures that occur when drilling glass fiber-reinforced composite are not high enough to make the protective action of the coating appreciable. Moreover, the effect of cutting speed on the wear of the high-speed steel drill can be seen clearly in Figure 3.4a: the farther from the drill center, the greater the loss of material on the major cutting edge.

In order to reduce tool wear when drilling glass fiber-reinforced epoxy composite laminates, low frequency and high amplitude axial vibration were induced in the feed direction [ARU 06]. As a result, tool wear was reduced in comparison with conventional drilling, probably owing to the reduction of both the coefficient of friction and temperature in the cutting zone.

(a) High-speed steel

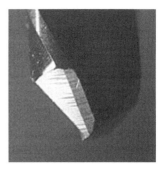

(b) Cemented carbide

(c) TiN coated carbide

Figure 3.4. *Tool wear after drilling 1,000 holes in glass fiber-reinforced epoxy composite using the following tool materials: (a) high-speed steel, (b) cemented carbide and (c) titanium nitride coated carbide*

The influence of feed rate (f) and cutting speed (v_c) on tool wear after drilling 1,000 holes in glass fiber-reinforced epoxy composite laminates using high-speed steel can be observed in Figure 3.5. Increasing feed rate from 0.04 to 0.2 mm/rot results in a reduction in tool wear, irrespectively of the cutting speed employed, owing to the reduction in the effective contact length between the tool and the abrasive glass fibers. However, when the cutting speed is increased from 55 to 86 m/min, its effect on tool wear cannot be clearly noticed, although we could expect an increase in tool wear as cutting speed is elevated, especially in the case of the high-speed steel, the strength of which is sensitive to a minimal increase in cutting temperature. Under the same conditions, negligible wear was observed on tungsten carbide drills.

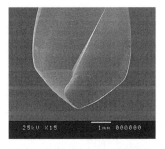 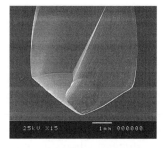

(a) v_c=55 m/min and f=0.04 mm/rot   *(b) v_c=55 m/min and f=0.2 mm/rot*

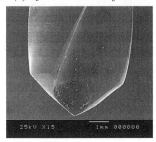 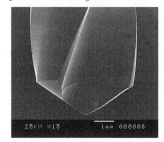

(c) v_c=86 m/min and f=0.04 mm/ rot   *(d) v_c=86 m/min and f=0.2 mm/ rot*

Figure 3.5. *Influence of feed rate and cutting speed on tool wear of high-speed steel: (a) v_c=55 m/min and f=0.04 mm/rot, (b) v_c=55 m/min and f=0.2 mm/rot, (c) v_c=86 m/min and f=0.04 mm/rot and (d) v_c=86 m/min and f=0.2 mm/rot*

In contrast, [ARU 07] reports that the wear of high-speed steel drills increases with feed rate when drilling glass fiber-reinforced epoxy composite laminates. As far as the cutting speed is concerned, however, a clear trend was not evident.

The influence of cutting speed on tool wear when drilling carbon fiber-reinforced polymeric laminates is reported by [CHE 97]. In addition to this, the author measured the temperature at the clearance face of the drill using two embedded thermocouples. The findings indicated that the clearance face temperature increased with cutting speed (reaching a maximum value of approximately 300°C at a cutting speed of $v_c = 200$ m/min) and decreased as the feed rate was elevated. Increasing the feed rate results in shorter contact length and, consequently, lower temperature.

The effect of the machining parameters on the temperature in the cutting zone (measured with an optical pyrometer) when face turning a carbon fiber-reinforced phenolic composite with PCD and titanium nitride-coated tungsten carbide inserts was reported by [SRE 99]. The temperature in the cutting zone increased with cutting speed, feed rate and depth of cut for both tool materials, although lower temperatures were recorded for the PCD. However, this elevation in temperature was steeper when the cutting speed exceeded 100 m/min for the coated carbide and 300 m/min for the PCD. As far as the wear mechanisms are concerned, the coated carbide tools failed due to abrasion with the exposure of the substrate, whereas the PCD tool failed by chipping. The superior hardness of PCD resulted in lower wear rates which, together with its high thermal conductivity, promoted lower machining temperatures.

The effect of tool wear on the compressive strength and fatigue life of carbon fiber-reinforced epoxy coupons was investigated by [PER 97], who drilled holes in specimens prior to static and fatigue testing using fresh and worn drills made of three distinct geometries and materials. The results indicated a reduction in the strength and fatigue life of all specimens that had holes produced with worn tools in comparison with fresh tools.

According to [BHA 98], the wear of high-speed steel drills when machining an aramid fiber-reinforced composite can be reduced by drilling under cryogenic temperatures and using negative point angle drills, in spite of the fact that the thrust force is elevated.

3.5. Drilling forces

Thrust force and torque developed when drilling fiber-reinforced polymeric composites are subjects of great interest owing to their influence on tool life and on the quality of the machined holes, especially on delamination. The prediction of thrust force and torque is not a straightforward task due to the anisotropy and non-homogeneity of these materials. As a consequence, the experimental approach is the one most frequently employed, although a considerable amount of work, e.g. [LAC 01], [LAN 05] and [PIR 05], has been concentrated on modeling forces and torque developed when machining composite materials.

A review on force models is presented by [GOR 03], although most theoretical works reported are specifically concerned with orthogonal cutting of composites reinforced with unidirectional fibers.

The behavior of thrust force and torque when drilling polymeric composite laminates with and without back support is described by [CAP 04]. When drilling using backing, initially the thrust force increases steeply as the chisel edge penetrates the work material and after that increases smoothly with large scatter as the cutting lips penetrate the material. This scatter is caused by the distinct properties of the matrix and reinforced materials. As the exit surface of the laminate approaches, delamination starts to take place and the thrust force decreases drastically.

The typical pattern for thrust force when drilling fiber-reinforced composites with back support is shown in Figure 3.6. When drilling without support, the thrust force increases linearly due to the laminate inflection. However, as the drill point approaches the exit side, a sudden decrease in the thrust force is observed.

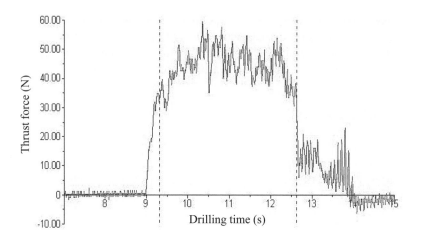

Figure 3.6. *Thrust force behavior when drilling fiber-reinforced composite*

In the case of torque, it increases to reach its maximum value when the cutting lips are fully engaged and decreases as the drill approaches the exit surface and delamination begins. The behavior of torque when drilling without backing is that of a sudden increase from zero to a peak value when the cutting lips fully engage the work material.

The influence of the matrix material and fiber shape on thrust force and torque was assessed by [KHA 04], who reported that when drilling woven-reinforced polymeric composites, the matrix material (epoxy and polyester) did not affect thrust force drastically, although the fiber shape presented a significant influence on thrust force, with higher values being observed when drilling the composite reinforced with chopped fibers in comparison with woven fibers, in spite of the lower volume fraction of the former reinforcing material.

As far as torque is concerned, higher values were recorded when drilling woven-reinforced polyester composite, the reason for that being the higher tangential force owing to the compressively loaded fibers.

Increasing the volume fraction of glass fibers from 0 to 23.7% promoted the elevation of both thrust force and torque owing to the enhancement of the mechanical properties (and cutting resistance) of the work material [ELS 04].

The effect of the cutting parameters on thrust force and torque when drilling high volume fraction (v_f = 66%) glass fiber-reinforced composite used for ballistic applications was investigated by [VEL 05]. Lower thrust force and torque values were obtained when cutting at a higher cutting speed and lower feed speed. Furthermore, tool wear increased gradually up to 300 holes and after that a drastic increase in the wear rate was observed.

The influence of feed rate on thrust force is consensual, i.e. an increase in feed rate promotes the elevation of the thrust force due to the increase in the shear plane area. In contrast, the influence of cutting speed is not yet well understood. According to [ARU 06], there is an optimal cutting speed value (v_c = 18.85 m/min when drilling glass fiber-reinforced epoxy resin) that leads to minimal thrust force.

Furthermore, the authors claim that vibratory drilling at low frequency and high amplitude (200 Hz and 15 μm, respectively) results in lower thrust force due to the cutting energy concentration, which reduces chip deformation.

Figure 3.7 shows the influence of feed rate and cutting speed on thrust force when drilling glass fiber-reinforced composite using tungsten carbide and high-speed steel helical drills. These results show that thrust force is considerably affected by feed rate.

In contrast, increasing cutting speed does not lead to any significant alteration on thrust force, except when using the tungsten carbide drill at a feed rate of f = 0.1 mm/rot.

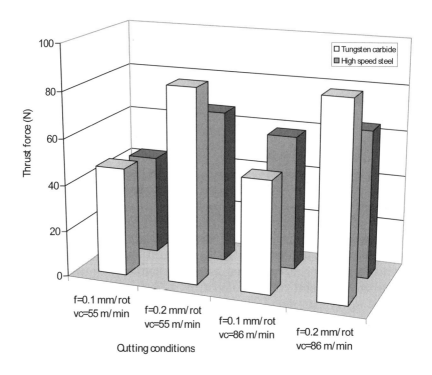

Figure 3.7. *Effect of cutting speed and feed rate on thrust force when drilling glass fiber-reinforced epoxy composite*

When drilling unidirectional carbon fiber-reinforced composite with high-speed steel and cemented carbide drills, [CHE 97] found that the thrust force and torque increased with feed rate, while the influence of cutting speed was negligible. High-speed (9,550-38,650 rpm) drilling tests using the same composite as work material and carbide drills indicated that the thrust force increased with cutting speed, even though the reason for such elevation was the accelerated tool wear rates, which drastically altered the geometry of the cutting edge [LIN 96].

Furthermore, the negative point angle drill was more prone to tool wear than the standard point twist drill and, consequently, promoted higher thrust force values. Finally, torque increased with cutting speed for the negative point angle drill and decreased for the twist drill.

In contrast, [ELS 04] noticed that when drilling epoxy resin reinforced with glass fibers at various volume fractions using high-speed steel drills, thrust force and torque decreased when the cutting speed was elevated. Similar results were reported by [KHA 04] who asserted that the reduction in thrust force and torque as cutting speed is elevated may be due to the softening of the matrix caused by heat generation assisted by the low coefficient of thermal conduction and low glass transition temperature of plastics.

With regard to the influence of the drill geometry on thrust force and torque, the point angle presents opposite effects on thrust force and torque [CHE 97]. An increase in the drill point angle leads to the elevation of the former and reduction of the latter. The increase in thrust force as the drill point angle is elevated is attributed to the increase in the shear area normal to feed direction. The author claims that the reduction in torque is caused by the elevation of the orthogonal rake angle with the drill point angle. Moreover, thrust force and torque decrease as the helix angle is increased and a similar behavior was observed when the chisel edge angle was elevated. Finally, higher thrust force and torque values were recorded when the web thickness was increased.

The influence of the drill point angle was also investigated by [ENE 01], who tested five distinct point angle values ranging from 75 to 160° when drilling carbon fiber-reinforced epoxy resin. Using Taguchi's robust design approach, the authors found that the lowest thrust force was obtained with the smallest tool point angle. Additionally, the influence of cutting speed and feed rate was investigated and the results indicated that thrust force decreased as cutting speed was elevated and increased with feed rate.

The performance of four carbide drills with different geometries (step drill, eight facet drill, four facet drill and parabolic point drill) used to machine unidirectional glass fiber-reinforced epoxy composite laminates indicated that the four facet drill, which is similar to a standard point twist drill with an additional clearance face, promoted the highest force and torque values, whereas the eight facet drill promoted lowest thrust force and the lowest step drill torque [SIN 06]. Additionally, thrust force decreased as cutting speed was elevated and increased with feed speed, while torque remained unaltered as cutting speed was elevated and increased slightly with feed speed.

An experimental study on the behavior of two cemented carbide drills with distinct geometries (helical point and negative point angle drills) when drilling polyester matrix reinforced with glass fibers was conducted by [DAV 04a]. The results indicated that the specific cutting force, which is directly dependent on torque and inversely dependent on shear area, decreased as feed rate was elevated, but it was not significantly affected by cutting speed. As far as the thrust force is concerned, it increased with feed rate and was not affected by cutting speed at a significant level of 5%. In addition to that, the negative point angle drill provided thrust force values substantially lower compared with the helical point drill.

Interestingly, opposite results were found when the same drills were tested against carbon fiber-reinforced epoxy laminates, i.e. the helical drill promoted lower specific cutting force and power than the negative point angle drill when drilling under the same cutting conditions [DAV 03a]. Additionally, drilling power increased with both cutting speed and feed rate. These findings suggested that the increase in cutting temperature due to the elevation in cutting speed has a negligible influence on reducing the shear strength of the work material, probably owing to the fact that the temperature is high enough to affect the strength of the matrix, but not of the reinforcing material.

An analytical and experimental investigation on the influence of the drill geometry on thrust force and delamination was carried out by Hocheng and Tsao [HOC 06]. The performance of the following high-speed steel drills when machining carbon fiber-reinforced laminates was assessed: standard twist drill, saw drill, candle stick (negative point angle) drill and step drill. The results indicated that the highest thrust force values were obtained using the twist drill, especially at higher feed rates, followed by the saw and core drills. The lowest thrust forces and, consequently, less delamination on the composite, were recorded for the step and negative point angle drills. Finally, the authors asserted that in order to reduce delamination, the drill geometry must allow the distribution of the thrust force along the drill periphery.

An analysis of the thrust force and torque obtained when drilling unidirectional glass fiber-reinforced epoxy resin with a high-speed steel trepanning tool indicated that the peak values for thrust force were found to be half of the values recorded when using the twist drill, together with slightly lower torque [MAT 99]. Reduced thrust force and torque were obtained by Piquet *et al.* [PIQ 00] using a specially designed drill with three cutting edges and a variable cutting edge angle (59°-0°).

According to these authors, torque is the result of the sum of two components: one related to the action of the primary cutting edges and the other as a result of friction between the secondary cutting edges and the hole wall, thus, using a variable cutting edge angle leads to a reduction in the contact area and, consequently, in the friction component of torque.

Figure 3.8 shows the influence of drill geometry on thrust force when drilling glass fiber-reinforced epoxy resin at various cutting conditions. Comparable force values were observed for the helical and negative point angle drills (see Figures 3.2a and 3.2e, respectively) at a lower feed rate (f = 0.1 mm/rot), owing to the lower point angle of 118° and thinned web of the helical drill.

Nevertheless, when feed rate was elevated to f = 0.2 mm/rot, the helical drill outperformed the negative point angle drill. The highest thrust force values were recorded when using the three flute drill (Figure 3.2d). Increasing the number of cutting edges we would expect a reduction in thrust force, but this was not true for the three flute drill owing to its large point angle. Similar to the results presented in Figure 3.7, thrust force increased with feed rate but it was not drastically affected by cutting speed.

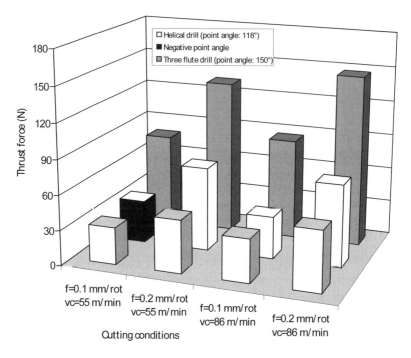

Figure 3.8. *Effect of tool geometry on thrust force when drilling glass fiber-reinforced epoxy composite*

A reduction in thrust force, torque and tool wear when drilling glass fiber-reinforced epoxy resin laminates with high-speed steel tools can be obtained by inducing tool vibration in the feed direction at a frequency of 220 Hz [RAM 02], owing to the fact that, in contrast to conventional drilling where force fluctuation takes place as the drill lips encounter two distinct materials, lower force fluctuation is observed for oscillatory assisted drilling. Not surprisingly, increasing drill diameter and keeping the rotational speed constant resulted in higher thrust force and torque values, as reported by [DHA 00] and [ELS 04].

The influence of tool wear on thrust force when drilling glass fiber-reinforced epoxy resin using tungsten carbide and high-speed steel helical drills is shown in Figure 3.9. While the thrust force required to drill with tungsten carbide increases smoothly with the number of holes reaching 135 N after 20,000 holes, the thrust force for the high-speed steel increased from 59 N in the first hole to 454 N when the thousandth hole is drilled.

These results indicate that the lower wear resistance of high-speed steel leads to high wear rates, probably caused by abrasion, which alter the tool tip geometry and promote an increase in thrust force. Furthermore, the temperature developed at the cutting zone may accelerate this process.

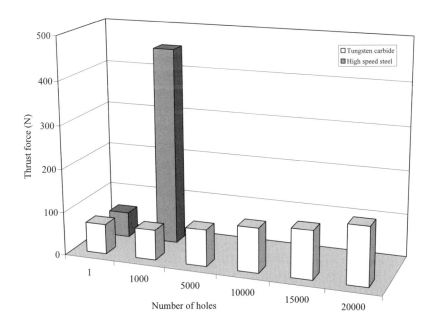

Figure 3.9. *Effect of tool wear on thrust force when drilling a glass fiber-reinforced epoxy composite*

3.6. Surface integrity

Surface integrity is the term used to describe the requirements that a machined component must satisfy in terms of its surface texture and metallurgy. Although polymeric materials reinforced with fibers are obviously not subjected to metallurgical alterations, their surface integrity can be drastically affected by the machining conditions. Furthermore, the glass transition temperature for many resins used as matrix materials is relatively low and reached easily when cutting. In the case of epoxy resin, the glass transition temperature is 150°C [ELS 04].

In general, holes produced in components for aerospace and automotive components are responsible for transferring loads within the structure and thus the quality and accuracy of the hole is critical to the strength of the joint [PER 97]. In the particular case of the aircraft industry, the economic impact of the damage induced by drilling is significant, especially when considering the added value associated with the component when it reaches the assembly stage [STO 96].

Ghidossi *et al.* [GHI 06] report that the machining conditions alter the tensile failure stress of glass fiber-reinforced epoxy composites. The influence of defects generated after making holes on the strength and fatigue life of composite laminates has been investigated by [PER 97], [RAM 02] and [TAG 90]. Generally speaking, the results indicated that tool material and geometry affect the fatigue life of composites slightly.

Four principal types of damage can take place when drilling fiber-reinforced polymeric composites [LAC 01]: delamination at the drill entrance (peel-up delamination), typically observed in unidirectional plies and leading to the tearing of the first ply; geometric defects, caused by alternated bending/compression stresses; thermal damage due to friction between fibers and the minor cutting edges; and delamination at the drill exit (push-up delamination), which is generally larger than at the drill entrance due to the fact that the uncut material cannot withstand the normal stress, leading to cracks. Nevertheless, most authors agree that delamination is regarded as the principal and most deleterious damage that can be induced when machining reinforced polymeric composites, as reported by [ZHA 01] and [GHI 04].

According to [KHA 04], delamination at the drill entrance is caused by the cutting force action, which generates a peeling force in the axial direction through the drill flute and resulting in the separation of laminas. With regard to the push-up delamination, as the drill approaches the exit side the uncut thickness is reduced leading to a reduction in the resistance to deformation. When the loading exceeds the interlaminar bond strength, delamination takes place.

3.6.1. *Delamination*

Various approaches have been proposed in order to minimize or eliminate delamination when drilling composites and the following approaches are noteworthy: drilling with a back support and the use of variable feed speed systems. The former is the simplest way to minimize delamination. Although machining time is not altered, production rate is reduced due to the need to bind and remove the support, before and after machining respectively. As an alternative to keep production rate unaltered, two or more laminates can be drilled simultaneously as long as the geometry of the parts allows. In this case, the outer laminates would act as back support.

Mechanical devices were initially used as damping systems aiming for progressive reduction in feed speed, thus minimizing delamination. The use of a viscous damping device is reported by [CAP 04] to reduce the delamination area to values comparable to those obtained when drilling with a support. With the advent of computer numerical control machines, this task has become commonplace.

An alternative approach proposed by [STO 96] consists of using a thrust force controller based on artificial neural networks capable of selecting the feed rate which will lead to the lowest thrust force and, consequently, to minimal damage when drilling graphite-epoxy laminates with diamond-tipped drills. Online monitoring of the thrust force has been employed by [ARU 06] in order to produce damage-free holes in glass fiber-reinforced epoxy laminates.

The assessment of the damage induced in polymeric composites is seldom carried out using destructive techniques. A metallographic technique (cross-sectioning followed by polishing and observation under the optical microscope) was performed by [CAP 95]. Despite being time consuming, the analysis of samples obtained at various drilling depths allowed the damage history to be depicted. Distinct failure modes were observed by the authors: delamination perpendicular to the feed direction and propagating within the matrix, step-like delamination, oblique fractures generated by the drill lips and high density microfailures.

In contrast, non-destructive techniques based on visual inspection are widely employed. A charge-coupled device (digital camera) attached to the toolmaker's microscope and connected to a computer with image processing software is the typical equipment used for the identification and measurement of the damage features. A scanner is also used for image acquisition.

In recent years, more sophisticated non-destructive techniques have been developed and applied to the detection and measurement of internal damage induced during the manufacturing and machining of composite materials. The use of high-speed speckle interferometry applied to the detection of subsurface delamination defects in carbon fiber composite is reported by [DAV 03c]. Ultrasonic scanning and X-ray tomography are also successfully employed for damage monitoring [HOC 06].

A method used to measure delamination based on the shadow moiré laser technique is presented by [SEI 07]. The results showed that the delamination area is larger at the drill exit and the damage increases with feed rate and is reduced as the rotational speed is elevated. The application of a fluorescent dye that penetrates into the delamination zone and can be observed under ultraviolet light is reported by Singh and Bhatnagar [SIN 06].

The importance of the use of inline non-destructive inspection techniques to automatically assess the quality of reinforced composites is discussed by [SCH 05]. According to the authors, ultrasonic inspection systems are capable of detecting internal defects such as voids, delamination, inclusions and cracks. In addition to this, machine vision systems can be developed to automatically inspect texture, position and geometry of the reinforcing material and impregnation quality, during the entire production of the composite laminates.

Nevertheless, the parameters employed to assess the induced damage may differ considerably and the terminology used is not uniform. Various dimensional and non-dimensional parameters are used to quantify the damage induced in polymeric composites. Among the dimensional parameters are: the difference between the damage area and area corresponding to the drill diameter; the difference between the maximum damage radius (or diameter) and the drill radius (or diameter); and the damage radius. Additionally, gobal damage (L_c) is the term used by [CAP 95] and it refers to the sum of the lengths of all fractures observed in cross-section maps of the hole.

Non-dimensional parameters are broadly used probably because they allow comparisons between tools with various diameters. The delamination factor (F_d) is the preferred parameter used to characterize the level of damage at the entrance and exit of the drill. It is calculated as the ratio of the maximum diameter of the delamination zone to the drill diameter. Similarly, the ratio of the damage area to the area corresponding to the drill diameter (or the ratio of the drill diameter to the maximum delamination diameter) is employed as a parameter to evaluate delamination.

According to [DAV 07], the delamination factor provides satisfactory results when delamination possesses a regular pattern, such as that observed in glass fiber-reinforced plastics. Nevertheless, when drilling carbon fiber-reinforced composites, delamination presents an irregular form, containing breaks and cracks at the hole entrance and exit.

In this case, the conventional delamination factor is not appropriated due to the fact that the size of the crack is not a convenient representation of the damage magnitude. Therefore, a parameter devised to take into account the contributions of the crack size (conventional delamination factor, F_d) and the damage area is proposed, i.e. adjusted delamination factor (F_{da}). Thus, if the trend is a delamination area equal to the crown area of maximum diameter (D_{max}) of the delamination zone, the adjusted delamination factor (F_{da}) presents a value equal to the square of the conventional delamination factor (uniform behavior).

If, however, the delamination area is minimal, the adjusted delamination factor (F_{da}) presents a value tending to the conventional delamination factor. Figure 3.10 presents a schematic diagram and photograph of the damages induced after drilling fiber-reinforced polymeric composites and selected parameters used to quantify the damage.

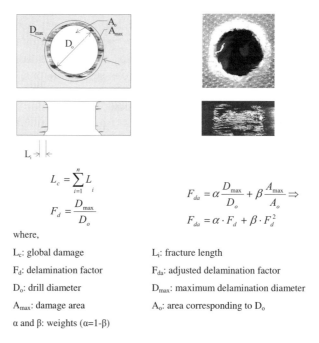

$$L_c = \sum_{i=1}^{n} L_i$$

$$F_d = \frac{D_{max}}{D_o}$$

$$F_{da} = \alpha \frac{D_{max}}{D_o} + \beta \frac{A_{max}}{A_o} \Rightarrow$$

$$F_{da} = \alpha \cdot F_d + \beta \cdot F_d^2$$

where,

L_c: global damage
F_d: delamination factor
D_o: drill diameter
A_{max}: damage area
α and β: weights ($\alpha = 1 - \beta$)

L_i: fracture length
F_{da}: adjusted delamination factor
D_{max}: maximum delamination diameter
A_o: area corresponding to D_o

Figure 3.10. *Schematic diagram and photograph of the damages induced by drilling reinforced polymeric composites and selected evaluation parameters*

As tool wear progresses and higher thrust force and torque are required to shear the material, the damage zone around the holes increases; see [AOY 95], [BHA 98] and [CHE 97]. These authors asserted that the relationship between the delamination factor and average thrust force is distinct when different materials are machined using the same tool cutting parameters. In addition to that, delamination at the drill exit is higher than at the drill entrance.

A comparison between the damage induced in distinct composite materials (chopped glass fiber-reinforced polyester, woven glass fiber-reinforced polyester and woven glass fiber-reinforced epoxy) after drilling using standard high-speed steels drills was carried out by [KHA 04]. In spite of requiring the highest thrust force, the composite reinforced with chopped fibers presented the least delamination, followed by the woven glass fiber-reinforced epoxy composite. According to the author, the woven glass fiber-reinforced polyester composite presented the highest level of delamination owing to its low fiber-matrix interface bond strength.

As far as the cutting parameters are concerned, feed rate is believed to be the principal factor affecting delamination. According to [CAP 04], under low feed rates delamination does not take place, however, when feed rate is increased the actual back rake angle becomes negative, thus pushing the work material instead of shearing and causing its delamination. The use of drills with geometry especially designed to machine fiber-reinforced polymeric composites may lead to a higher threshold feed rate required to induced delamination at drill entrance [HOC 06]. Piquet *et al.* [PIQ 00] asserts that the specially designed drills allowed a significant reduction in the damage compared to helical point drills.

The damage area observed when drilling carbon fiber-reinforced epoxy laminates increases with feed rate and decreases as cutting speed was elevated [SEI 07]. An experimental study on the drilling of carbon fiber-reinforced composites carried out by [DAV 03a] suggests that delamination at the drill entrance and exit are affected by distinct parameters, i.e. feed rate is the principal factor affecting delamination at the drill entrance and cutting speed is the main factor affecting delamination at the drill exit. Furthermore, using a carbide helical drill, the delamination factor increases with cutting speed and feed rate, while the influence of the cutting parameters is nearly negligible when a negative point angle drill is tested.

An analysis of variance was conducted by [TSA 04] in order to evaluate the statistical influence of cutting speed, feed rate and tool diameter on the delamination factor when drilling carbon fiber-reinforced epoxy resin laminates using three high-speed steel drill geometries: standard twist drill, negative point angle drill and saw drill. The findings indicated that when the twist drill was tested, the delamination factor was affected by feed rate only, whereas drill diameter was the only factor which significantly affected the delamination factor when the negative point angle drill was employed. With regard to the saw drill, the cutting speed, feed rate and drill diameter significantly affected the delamination factor within the tested range. In addition to this, the negative point angle and saw drills produced lower delamination factor values in comparison with the twist drill.

Figures 3.11 and 3.12 show the influence of feed speed (v_f) and tool geometry on the delamination factor when drilling carbon fiber reinforced epoxy laminates with tungsten carbide drills at cutting speeds of $v_c = 63$ and 126 m/min, respectively. Three tool geometries were tested: two helical drills with point angles of 85° and 115° and one negative point angle drill. Observing these figures it can be noted that the delamination factor increases with feed speed, particularly in the case of the helical point drills. When the negative point angle drill is tested, the delamination factor remains unaltered at higher feed speeds. Furthermore, the helical drill with the largest point angle is responsible for the highest delamination factor values, irrespective of the cutting speed and feed speed employed. As previously discussed, this behavior can be explained by the highest thrust force expected when the point angle is elevated. Finally, these figures indicate that comparable values for the delamination factor are obtained when drilling with the helical drill with a point angle of 85° and the negative point angle drill at feed speeds up to $v_f = 6,000$ mm/min

Comparing Figures 3.11 and 3.12, the influence of cutting speed on the delamination factor can be estimated, i.e. in general, the delamination factor decreases as cutting speed is elevated.

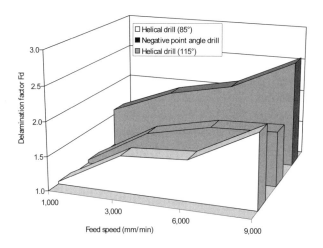

Figure 3.11. *Effect of feed speed and drill geometry on delamination factor when drilling carbon fiber-reinforced epoxy composite at $v_c = 63$ m/min*

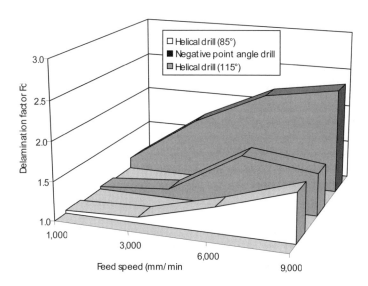

Figure 3.12. *Effect of feed speed and drill geometry on delamination factor when drilling carbon fiber-reinforced epoxy composite at $v_c = 126$ m/min*

The introduction of high-speed machining has led to an appreciable increase in productivity (material removal rate), thus reducing machining costs, although high-speed drilling of fiber-reinforced composites has not received as much attention as other operations, such as high-speed milling of dies and moulds for example. The chips produced by the drilling of carbon composites are rather abrasive and cause high tool wear rates, thus increasing the required thrust force and inducing damage [FER 06].

When high-speed drilling (rotational speeds ranging from 20,000 to 80,000 rpm and feed rates from 10 to 50 μm/rot) small diameter holes (Ø=0.4–5 mm) with tungsten carbide drills, [INO 97] noted that the ratio of the drill radius to the yarn width significantly affects tool life. In addition to that, higher tool wear rates were observed when drilling at low cutting speed and feed rate values and using drills of larger diameter. Nevertheless, it is important to point out that only when investigating the effect of feed rate, tool wear was measured directly on the drills. When studying the influence of cutting speed and drill diameter, tool wear was assessed in terms of hole diameter, which may mislead the results.

An investigation into the influence of drill geometry and spindle speed (ranging from 9,550 to 38,650 rpm) on the hole quality of carbon fiber-reinforced laminates was conducted by [LIN 96], who noticed that increasing cutting speed results in lower thrust force values, which is closely related to lower delamination levels. Nevertheless, tool wear is the major limiting factor when high-speed drilling carbon fiber composites due to its deleterious influence on thrust force. Comparing the performance of uncoated and coated drills when drilling composites, [KAO 05] asserts that a two-fold increase in tool life can be obtained by using MoS_2-Cr-coated drills, which were able to produce holes with high quality using rotational speeds up to 100,000 rpm.

Similarly, work by [CAM08] involving high-speed drilling of glass reinforced epoxy laminates with two carbide helical point drills (point angles of 85° and 115°) and one carbide negative point angle drill using spindle speeds from 4,000 to 40,000 rpm indicated that delamination decreases as cutting speed is elevated, probably owing to the softening of the matrix caused by higher temperatures. Using lower spindle speeds (4,000 and 8,000 rpm), delamination increased with feed speed, but the damage was not affected by feed speed when a spindle speed of 40,000 rpm was employed. Finally, the negative point angle drill was responsible for the least delamination at the highest spindle speed; however, it was outperformed by the helical drill with a point angle of 85° under lower spindle speeds.

3.6.2. *Surface roughness*

Although it is not clear whether 2D surface roughness parameters can provide an accurate estimate of the condition of the machined surface, the use of parameters such as R_a, R_q, R_y and R_z to describe polymeric composite surfaces is widespread due to the fact that they have been successfully applied to represent the texture of metals and are readily obtained using portable equipment. Nevertheless, in contrast to metallic alloys, which possess a relatively rigid and continuous surface, fiber-reinforced polymeric composites may present defects induced by machining operations, which cannot be characterized using a contact probe such as a roughness meter stylus.

The influence of the surface finish on the mechanical properties of polymeric composites was investigated by [ERI 00]. Specimens made of short glass fiber-reinforced thermoplastics were face turned in order to present three distinct roughness levels.

Impact, bending and fatigue tests were carried out and the results indicated that these properties were not significantly affected by the surface roughness of the samples, probably owing to the weak bonding between the fibers and matrices used in this study. Similar results were reported by [GHI 04], who investigated the influence of the surface roughness of milled specimens on mechanical properties. However, according to the authors, surface roughness is not a suitable parameter to represent the damage induced by the machining of composites.

In general, high cutting speeds and low feed rates are recommended in order to produce holes with good surface finish in reinforced polymeric composites [ENE 01]. Due to the low thermal conductivity of these materials, surface roughness increases as cutting speed is elevated and feed rate is reduced owing to the heat built up. Similarly to the machining of metals, feed rate is the most relevant factor affecting roughness of the hole wall [OGA 97]. The reason for this resides in the fact that the distance between peaks and valleys decreases as the feed rate is reduced. Moreover, using lower feed rates the thrust force is reduced, thus improving the quality of the machined surface.

The surface roughness obtained on machined reinforced polymeric composites is strongly dependent on fiber orientation, i.e. lower roughness is obtained when cutting parallel to the fiber direction. Furthermore, when drilling a composite reinforced with woven cloth, the surface roughness on the hole wall is variable, depending on the cutting edge position angle, i.e. maximum peak to valley roughness values are observed at an edge position angle of approximately 30° [AYO 95].

When drilling graphite-bismaleimide stacks, Kim and Ramulu [KIM 04] noted that the surface damage (fiber pullout) generally occurred with the fiber at a negative angle to the cutting direction. In addition to this, the maximum peak to valley roughness increased with feed rate and the high-speed steel drill provided higher values compared with the carbide tool.

Spindle speed was the principal factor affecting the roughness of holes produced by the steel tool and feed rate was the principal factor affecting roughness when drilling with the carbide drill. The reason for that is attributed to the fact that, in contrast with the carbide drill, the high-speed steel tool possesses inferior wear resistance, especially at higher speeds (higher temperatures in the cutting zone and longer contact length between the cutting edges and fibers).

Drilling trials on glass fiber-reinforced epoxy resin with various volume fractions (from 8.9 to 23.7%) using a standard point high-speed steel drill were carried out by [ELS 04]. Lower average roughness (R_a) values were obtained by increasing the cutting speed and fiber volume fraction. Increasing the cutting speed results in lower thrust force and, consequently, better surface finish. With regard to the fiber volume fraction, increasing the fibers percentage results in superior stiffness, higher thermal conductivity and higher glass transition temperature for the composite. Finally, the influence of feed rate on the roughness of the machined surface is affected by the fiber volume fraction, i.e. when drilling the composite with lower fiber volume fraction, higher surface roughness is recorded at the lower feed rate. In contrast, when the volume fraction is increased, the elevation in feed rate results in higher surface roughness values.

The influence of drill geometry (twist drill against negative point angle drill) and cooling environment (dry cutting against cryogenic cooling) on surface roughness when drilling aramid fiber-reinforced composite laminates was studied by [BHA 98]. Considerably lower roughness values were obtained under cryogenic cooling, irrespective of the drill geometry tested, owing to the fact that the cryogenic temperature induced higher stiffness and compressive stress caused by the difference between the coefficients of thermal expansion of the matrix and reinforcing.

Figure 3.13 shows the influence of the cutting parameters (cutting speed and feed rate) on the surface roughness of holes produced in glass fiber-reinforced epoxy resin laminates. In spite of the fact that lower R_a values were obtained employing the lower feed rate, this figure indicates that a direct relationship between the cutting parameters and surface roughness is not evident.

As far as the cutting speed is concerned, an increase may lead, on the one hand, to the reduction of the thrust force caused by the softening of the work material, thus improving surface finish. On the other hand, higher rotational spindle speeds are required in order to achieve higher tangential cutting speeds. Consequently, vibration may take place promoting a poorer surface finish. These findings suggest that the influence of cutting speed and feed rate on surface roughness is restricted and that the roughness values obtained should be used for general guidance only.

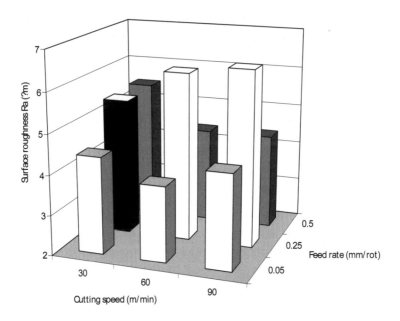

Figure 3.13. *Effect of cutting speed and feed rate on the average surface roughness (R_a) of glass fiber-reinforced epoxy composite*

Davim *et al.* [DAV 04a] investigated the influence of drill geometry on surface roughness of a glass fiber-reinforced polyester composite and found out that the negative point angle drill was responsible for superior finish (lower R_a values) compared with the helical point drill. Further work focused on the effect of the drilling parameters on the surface roughness of two glass fiber-reinforced polyester composites has been conducted [DAV 04b]. The analysis of variance indicated that within the cutting range tested, cutting speed and feed rate significantly affected surface roughness (R_a) and that the percentage of contribution of cutting speed was superior to that of feed rate. Surface roughness increased with feed rate and decreased as cutting speed was elevated. Similar results were reported by [TSA 08], who noticed that the surface roughness of holes drilled with a high-speed steel drill (negative point angle geometry) in carbon fiber-reinforced epoxy laminates were significantly affected by cutting speed and feed rate, but not by drill diameter.

3.7. Dimensional and geometric deviations

Dimensional accuracy of cylindrical holes produced in reinforced polymeric composites is focused essentially on hole diameter owing to the fact that blind holes are rarely produced and thus hole depth is not a major concern. As far as the geometric tolerances are concerned, form deviations, such as circularity (roundness) and cylindricity, and location deviation (position) are the principal subject matter.

In spite of the fact that they play a critical role in the performance of the part, the dimensional and geometric deviations obtained after drilling fiber-reinforced have not received the same attention as delamination or surface roughness. The need for studying dimensional and geometric deviations is evident when considering, for instance, the distinct levels of requirements for aircraft assembly: while the hole tolerance for riveting may be as high as 0.25 mm, a typical tolerance value for straight shank fasteners is ±13 µm [EVE 99]. In the case of bolt holes in the structure of an aeroplane wing and tail, diameter tolerances of 30 µm or less are required [BRI 02].

The tolerances obtained after machining are closely related to the cutting forces, i.e. the factors which lead to lower thrust force and torque will inevitably promote tighter deviations. Therefore, tool geometry and the cutting parameters, especially feed rate, must receive special attention. When drilling carbon fiber composite, tighter diameter tolerances are obtained using a step drill and minimal quantity lubrication results in comparison, respectively, with a standard point twist drill and dry cutting [BRI 02]. Trepanning tools are considered superior to twist drills [MAT 99] when drilling glass fiber-reinforced composite laminates due to the fact that the former promotes lower thrust force and torque values. Nevertheless, the position tolerance may be critical due to the absence of the chisel edge in the trepanning tool.

Tool wear is another relevant factor affecting tolerances for two reasons: firstly, tool wear leads to a reduction in drill diameter and thus undersize holes may be generated; and secondly, the deterioration of the cutting edge and consequent loss of sharpness results in higher thrust force and torque, which will contribute to an impairment of the quality of the machined holes. The influence of tool wear on the quality of small holes (Ø1 mm) produced in glass fiber-reinforced epoxy laminates using a constant spindle speed of 60,000 rpm and feed rate of 0.05 mm/rev was investigated by [AYO 95]. Hole radius decreases as the number of holes produced increases due to tool wear. However, the damage radius remained unaltered, thus suggesting that the damage extent increased with tool wear.

The influence of cutting speed and feed rate on the dimensional deviation (hole diameter) of the thousandth hole produced in glass fiber-reinforced epoxy laminates using high-speed steel helical drill (Ø5 mm) is given in Figure 3.14. These findings suggest that closer tolerances are obtained using a lower cutting speed and higher feed rate, the reason for that being that friction between the fibers and the secondary cutting edges increases at higher speed and lower feed rate, as long as the contact length is longer. As a consequence, the hole diameter is affected by tool wear.

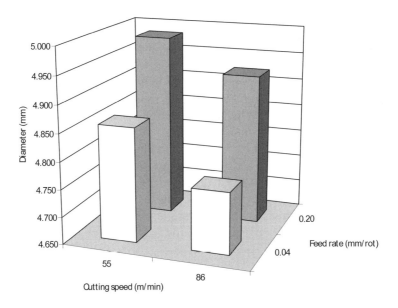

Figure 3.14. *Effect of cutting speed and feed rate on the dimensional deviation of holes produced in glass fiber-reinforced epoxy composite*

Hole shrinkage due to relaxation of the laminate after drilling glass fiber-reinforced epoxy resin using high-speed steel drills was assessed by [ARU 07]. Higher values for hole shrinkage were observed as the feed rate was elevated, probably owing to the progressive tool wear leading to higher order stressing of the material and consequent relaxation. As far as the cutting speed is concerned, a straightforward relationship with hole shrinkage was not detected, although the behavior of hole shrinkage and tool wear were quite similar as cutting speed was elevated.

In contrast, it has been reported by [VEL 05] that when drilling woven glass fiber-reinforced composites with a high fiber volume fraction, shrinkage does not take place due to the fact that the woven fabric does not allow the fiber to bend, thus minimizing relaxation. The tests results indicated an oversize ranging from 5 to 15 µm when drilling with a fresh carbide drill. As the number of holes produced (and tool wear) increased, oversize was reduced remaining stable from 200 to 400 holes. With regard to the cutting parameters, higher cutting speed and feed speed values resulted in lower oversize.

Dimensional accuracy of holes produced in aramid fiber-reinforced composite laminates can be improved considerably by using negative point angle drills and liquid nitrogen as cutting fluid [BHA 98]. Negligible tool wear was observed after drilling 250 holes with the negative point angle drill under cryogenic cooling. Consequently, surface finish and roundness error were not affected. The reason for this is attributed to the cutting mechanism imposed by the negative point angle drill (the fibers are subjected to tension towards the center of the material before shearing) and the low temperature effect previously discussed. In contrast, a roundness error ranging from 85 to 175 µm was observed when dry drilling 250 holes using a standard twist drill.

The performance of high-speed steel and carbide tools when drilling graphite/bismaleimide stacks was investigated by [KIM 04]. The findings indicated that the high-speed steel drill produced undersize holes, whereas oversize holes were generated by the carbide drill. Furthermore, the oversize increased with the elevation of rotational speed and feed rate, and higher roundness deviation values were observed for the high-speed steel tool. The elevation of rotational speed and the reduction of feed rate resulted in higher cylindricity deviations. Finally, the analysis of variance indicated that rotational speed was the principal factor affecting hole diameter and feed rate was the principal factor acting on cylindricity.

The circularity (roundness) deviation of the thousandth hole generated in glass fiber-reinforced epoxy laminates using high-speed steel helical drills (Ø5 mm) at various cutting conditions is shown in Figure 3.15. It can be noted that similar deviations were obtained irrespective of the cutting speed and feed rate employed. Furthermore, no relationship can be drawn with the dimensional deviation results presented in Figure 3.14, although we would expect that minimal circularity would be obtained under the machining condition responsible for tighter diametric deviation). These results suggest that, in addition to the machining conditions (cutting parameters, tool geometry and wear), the machine tool condition plays a significant role on the roundness results.

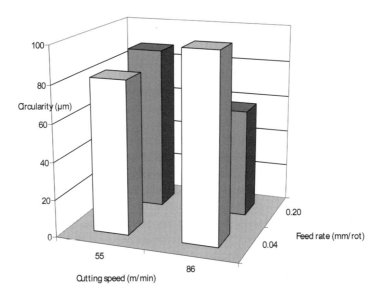

Figure 3.15. *Effect of cutting speed and feed rate on the circularity of holes drilled in glass fiber-reinforced epoxy composite*

3.8. Conclusions

Drilling of fiber-reinforced polymeric composites is a complex subject matter of increasing relevance due to the remarkable properties of these materials and the ever growing range of applications. Nevertheless, the cutting mechanisms involved are quite distinct from those observed when machining metallic alloys, thus requiring an in-depth analysis in order to optimize the operation in terms of both the performance of the cutting tool employed and the quality of the machined component. The principal characteristics of polymeric composite materials which affect their machinability are related to the reinforcing phase: fiber material, form, orientation and volume fraction.

High-speed steel and plain tungsten carbide are the principal materials used as cutting tools, prepared in a variety of geometries ranging from standard twist drills to tools specially designed to produce holes with the best possible quality. Particular attention is paid to the point angle and chisel length (web thickness), which seem to be the principal features affecting the performance of the drill. In spite of being quite suitable for drilling fiber-reinforced polymeric composites due to their superior wear resistance, diamond-tipped drills are seldom employed. Similarly, the use of coated high-speed steel and carbide tools is not widespread.

In addition to tool material and geometry, the cutting parameters (cutting speed and feed rate) directly affect production rate and cost, tool wear, drilling forces and the quality of the hole. However, in contrast to metals, the use of cutting parameters, which represent higher material removal rates, does not necessarily lead to higher tool wear rates and poorer quality of the machined component. For instance, tool wear appears to be reduced as feed rate is elevated and the damage around the hole produced by certain tool configurations is reduced when cutting speed is elevated. In order to minimize the damage induced as the drill exits the laminate, variable feed speed is frequently recommended, achieved by means of either mechanical devices in conventional drilling machines or through the numerical control of machine centers.

Abrasion caused by friction between fibers and the secondary cutting edges is the principal wear mechanism observed and chipping is frequently reported. The inferior wear resistance of high-speed steel results in accelerated wear rates, which alter the tool geometry by rounding the cutting edges. As a consequence, drilling forces increase and the quality of the hole is impaired.

The close relationship between drilling forces and both tool wear and the machined hole quality make thrust force and torque subjects of greatest interest. Due to the difficulties associated with measuring wear on drills, tool wear is frequently monitored through force signals. Not surprisingly, drilling forces increase with tool wear. Tooling and machining conditions responsible for increasing thrust force and torque will inevitably represent extended damage and inferior dimensional and geometric quality of the hole. In general, drilling forces increase with fiber volume fraction, drill point angle, chisel length and feed rate. Any alteration leading to an increase in the shear plane area results in higher thrust force, but the role of the cutting temperature is not completely understood: on the one hand, higher temperatures promote the softening of the matrix and the reduction of the strength of the composite, thus reducing thrust force. On the other hand, clogging of the matrix in the drill flutes may increase thrust force.

Delamination is regarded as the most critical damage induced after drilling reinforced polymeric composites as it can severely impair the performance of the component. Irrespective of the technique and parameter selected to assess the damage, the feed rate appears to be the principal factor affecting delamination. Tool geometry is another relevant factor and, in general, geometries which promote lower thrust force values induce less damage on the composite. In addition to this, less delamination is observed using drills that restrain the extrusion of the work material at the center of the hole by the chisel edge, such as the negative point angle drill, saw drill and trepanning tool.

Although extensively used to represent the machined surface texture of reinforced polymeric composites, standard 2D surface roughness parameters do not seem capable of accurately representing the condition of surfaces that may present damage not observed on metallic surfaces. Nonetheless, higher cutting speeds and lower feed rates are expected to produce a better surface finish. Surface roughness is further affected by fiber orientation; consequently, rather distinct roughness values can be obtained after drilling woven fiber-reinforced polymeric composites depending on the angular position where the stylus is placed.

In spite of their importance to the performance of the machined component, dimensional and geometric deviations obtained after drilling composites have not received as much attention as surface integrity. Diametric deviation, circularity and cylindricity are the principal parameters assessed and the required tolerance varies drastically depending on the application of the component. Generally speaking, tolerances are affected by drilling forces and so closer deviations are expected to be obtained after machining with tooling and cutting conditions which account for lower forces. Similarly, tighter deviations are obtained with tools with negligible wear. The condition of the machine tool is also significant to the dimensional and geometric deviations obtained.

3.9. Acknowledgements

The authors would like to thank CAPES-Brazil and FCT-Portugal for funding this research project. Additional thanks go out to Mr. Eduardo Antônio Carvalho, from CDTN, for his support with the scanning electron microscopy.

3.10. References

[ABR 07] ABRÃO A.M., FARIA P.E., CAMPOS RUBIO J.C., REIS P., DAVIM J.P., "Drilling of fiber reinforced plastics: a review", *Journal of Materials Processing Technology,* Vol. 186, 2007, p. 1–7.

[ADA 03] ADAMS R.D., MAHERI M.R., "Damping in advanced polymer–matrix composites", *Journal of Alloys and Compounds*, Vol. 355, 2004, p. 126–130.

[AN 97] AN S.-O., LEE E.-S., NOH S.-L., "A study on the cutting characteristics of glass fiber reinforced plastics with respect to tool materials and geometries", *Journal of Materials Processing Technology*, Vol. 68, 1997, p. 60-67.

[AOY 95] AOYAMA E., INOUE H., HIROGAKI T., NOBE H., KITAHARA Y., KATAYAMA T., "Study on small diameter drilling in GFRP", *Composite Structures*, Vol. 32, 1995, p. 567-573.

[ARU 06] ARUL S., VIJAYARAGHAVAN L., MALHOTRA S.K., KRISHNAMURTHY R., "The effect of vibratory drilling on hole quality in polymeric composites", *International Journal of Machine Tools and Manufacture*, Vol. 46, 2006, p. 252–259.

[ARU 07] ARUL S., VIJAYARAGHAVAN L., MALHOTRA S.K., "Online monitoring of acoustic emission for quality control in drilling of polymeric composites", *Journal of Materials Processing Technology*, Vol. 185, 2007, p. 184–190.

[BHA 95] BHATNAGAR N., RAMAKRISHNAN N., NAIK N.K., KOMANDURI R., "On the machining of fiber reinforced plastic (FRP) composite laminates", *International Journal of Machine Tools and Manufacture*, Vol. 35, no. 5, 1995, p. 701-716.

[BHA 98] BHATTACHARYYA D., HORRIGAN D.P.W., "A study of hole drilling in kevlar composites", *Composites Science and Technology*, Vol. 58, 1998, p. 267-283.

[BHA 04] BHATNAGAR N., SINGH I., NAYAK D., "Damage investigation in drilling of glass fiber reinforced plastic composite laminates", *Materials and Manufacturing Processes*, Vol. 19, no. 6, 2004, p. 995-1007.

[BOT 03] BOTELHO E.C., FIGIEL L., REZENDE M.C., LAUKE B., "Mechanical behavior of carbon fiber reinforced polyamide composites", *Composites Science and Technology*, Vol. 63, 2003, p. 1843–1855.

[BRI 02] BRINKSMEIER E., JANSSEN R., "Drilling of multi-layer composite materials consisting of carbon fiber reinforced plastics (CFRP), titanium and aluminum alloys", *CIRP Annals-Manufacturing Technology*, Vol. 51, no. 1, 2002, p. 87-90.

[CAM 08] CAMPOS RUBIO J., ABRÃO A.M., FARIA P.E., ESTEVES CORREIA A., DAVIM J.P., "Effects of high speed in the drilling of glass fibre reinforced plastic: evaluation of the delamination factor", *International Journal of Machine Tools and Manufacture*, Vol. 48, 2008, p. 715-720.

[CAP 95] CAPRINO G., TAGLIAFERRI V., "Damage development in drilling glass fibre reinforced plastics", *International Journal of Machine Tools and Manufacture*, Vol. 35, no. 6, 1995, p. 817-829.

[CAP 96] CAPRINO G., DE LORIO I., NELE L., SANTO L., "Effect of tool wear on cutting forces in the orthogonal cutting of unidirectional glass fibre-reinforced plastics", *Composites: Part A*, Vol. 27A, 1996, p. 409-415.

[CAP 04] CAPELLO E., "Workpiece damping and its effects on delamination damage in drilling thin composite laminates", *Journal of Materials Processing Technology*, Vol. 148, no. 2, 2004, p. 186-195.

[CHE 97] CHEN W.C., "Some experimental investigations in the drilling of carbon fiber-reinforced plastic (CFRP) composite laminates", *International Journal of Machine Tools and Manufacture*, Vol. 37, no. 8, 1997, p. 1097-1108.

[DAV 03a] DAVIM J.P., REIS P., "Drilling carbon fibre reinforced plastics manufactured by autoclave experimental and statistical study", *Materials and Design*, Vol. 24, no. 5, 2003, p. 315-324.

[DAV 03b] DAVIM J.P., REIS P., "Study of delamination in drilling carbon fiber reinforced plastics (CFRP) using design experiments", *Composite Structures*, Vol. 59, no. 4, 2003, p. 481-487.

[DAV 03c] DAVILA A., RUIZ P.D., KAUFMANN G.H., HUNTLEY J.M., "Measurement of sub-surface delaminations in carbon fibre composites using high-speed phase-shifted speckle interferometry and temporal phase unwrapping", *Optics and Lasers in Engineering*, Vol. 40, 2003, p. 447–458.

[DAV 04a] DAVIM J.P., REIS P., ANTÓNIO C.C., "Experimental study of drilling glass fibre reinforced plastics (GFRP) manufactured by hand lay-up", *Composites Science and Technology*, Vol. 64, no. 2, 2004, p. 289-297.

[DAV 04b] DAVIM J.P., REIS P., ANTÓNIO C.C., "Drilling fiber plastics (FRPs) manufactured by hand lay-up: influence of matrix (Viapal VUP 9731 and ATLAC 382-05)", *Journal of Materials Processing Technology*, Vol. 155–156, 2004, p. 1828–1833.

[DAV 07] DAVIM J.P., CAMPOS RUBIO J.C., ABRÃO A.M., "A novel approach based on digital image analysis to evaluate the delamination factor after drilling composite laminates", *Composites Science and Technology*, Vol. 67, 2007, p. 1939-1945.

[DHA 00] DHARAN C.K.H., WON M.S., "Machining parameters for an intelligent machining system for composite laminates", *International Journal of Machine Tools and Manufacture*, Vol. 40, 2000, p. 415–426.

[ELS 04] EL-SONBATY I., KHASHABA, U.A., MACHALY T., "Factors affecting the machinability of GFR/epoxy composites", *Composite Structures*, Vol. 63, no. 3-4, 2004, p. 329-338.

[ENE 01] ENEMUOH E.U., EL-GIZAWY A.S., OKAFOR A.C., "An approach for development of damage-free drilling of carbon fibre reinforced thermosets", *International Journal of Machine Tools and Manufacture*, Vol. 41, no. 12, 2001, p. 1795-1814.

[ERI 00] ERIKSEN E., "The influence of surface roughness on the mechanical strength properties of machined short-fibre-reinforced thermoplastics", *Composites Science and Technology*, Vol. 60, 2000, p. 107-113.

[EVE99] EVERSON C.E., CHERAGHI S.H., "The application of acoustic emission for precision drilling process monitoring", *International Journal of Machine Tools and Manufacture*, Vol. 39, 1999, p. 371–387.

[FAR 08] FARIA P.E., CAMPOS R.F., ABRÃO A.M., GODOY G.C.D., DAVIM J.P., "Thrust force and wear assessment when drilling glass fiber polymeric composite", *Journal of Composite Materials*, Vol. 42, no. 14, 2008, p. 1401-1414.

[FER 01] FERREIRA J.R., COPPINI N.L., LEVY NETO F., "Characteristics of carbon-carbon composite turning", *Journal of Materials Processing Technology*, Vol. 109, 2001, p. 65-71.

[FER 06] FERNANDES M., COOK C., "Drilling of carbon composites using a one shot drill bit. Part I: Five stage representation of drilling and factors", *International Journal of Machine Tools and Manufacture*, Vol. 46, 2006, p. 70-75.

[FU 00] FU S.-Y., LAUKE B., MÄDER E., YUE C.-Y., HU X., "Tensile properties of short-glass-fiber- and short-carbon-fiber-reinforced polypropylene composites", *Composites: Part A*, Vol. 31, 2000, p. 1117–1125.

[GHI 04] GHIDOSSI P., EL MANSORI M., PIERRON F., "Edge machining effects on the failure of polymer matrix composite coupons", *Composites: Part A*, Vol. 35, 2004, p. 989–999.

[GHI 06] GHIDOSSI P., EL MANSORI M., PIERRON F., "Influence of specimen preparation by machining on the failure of polymer matrix off-axis tensile coupons", *Composites Science and Technology*, Vol. 66, 2006, p. 1857–1872.

[GLI 05] GLIESCHE K., HÜBNER T., ORAWETZ H., "Investigations of in-plane shear properties of ±45°-carbon/epoxy composites using tensile testing and optical deformation analysis", *Composites Science and Technology*, Vol. 65, 2005, p. 163–171.

[GOR 03] GORDON S., HILLERY M.T., "A review of the cutting of composite materials", *Proceedings of the Institution of Mechanical Engineers, Part L: Journal Materials: Design and Applications*, Vol. 217, 2003, p. 35-45.

[HOC 06] HOCHENG H., TSAO C.C., "Effects of special drill bits on drilling-induced delamination of composite materials", *International Journal of Machine Tools and Manufacture*, Vol. 46, no. 12-13, 2006, p. 1403-1416.

[INO 97] INOUE H., AOYAMA E., HIROGAKI T., OGAWA K., MATUSHITA H., KITAHARA Y., KATAYAMA T., "Influence of tool wear on internal damage in small diameter drilling in GFRP", *Composite Structures*, Vol. 39, no. 1-2, 1997, p. 55-62.

[KAO 05] KAO W.H., "Tribological properties and high speed drilling application of MoS2-Cr coatings", *Wear*, Vol. 258, 2005, p. 812-825.

[KHA 04] KHASHABA U.A., "Delamination in Drilling GFR-Thermoset Composites", *Composite Structures*, Vol. 63, 2004, p. 313-327.

[KIM 04] KIM D., RAMULU M., "Drilling process optimization for graphite/bismaleimide–titanium alloy stacks", *Composite Structures*, Vol. 63, 2004, p. 101–114.

[LAC 01] LACHAUD F., PIQUET R., COLLOMBET F., SURCIN L., "Drilling of composite structures", *Composite Structures*, Vol. 52, no. 3-4, 2001, p. 511-516.

[LAN 05] LANGELLA A., NELE L., MAIO A., "A torque and thrust prediction model for drilling of composite materials", *Composites. Part A: Applied Science and Manufacturing*, Vol. 36, no. 1, 2005, p. 83-93.

[LIN 96] LIN S.C., CHEN I.K., "Drilling carbon fibre-reinforced composite material at high speed", *Wear*, Vol. 194, no. 1-2, 1996, p. 156-162.

[MAT 99] MATHEW J., RAMAKRISHNAN N., NAIK N.K., "Investigations into the effect of geometry of a trepanning tool on thrust and torque during drilling of GFRP composites", *Journal of Materials Processing Technology*, Vol. 91, no. 1, 1999, p. 1-11.

[OGA 97] OGAWA K., AOYAMA E., INOUE H., HIROGAKI T., NOBE H., KITAHARA Y., KATAYAMA T., GUNJIMA M., "Investigation on cutting mechanism in small diameter drilling for GFRP (thrust force and surface roughness at drilled hole wall)", *Composite Structures*, Vol. 38, no. 1-4, 1997, p. 343-350.

[PAR 03] PARK S.-J., SEO M.-K., SHIM H.-B., "Effect of fiber shapes on physical characteristics of non-circular carbon fibers-reinforced composites", *Materials Science and Engineering A*, Vol. 352, 2003, p. 34-39.

[PER 97] PERSSON E., ERIKSSON I., ZACKRISSON L., "Effects of hole machining defects on strength and fatigue life of composite laminates", *Composites Part A*, Vol. 28A, 1997, p. 141-151.

[PIR 05] PIRTINI M., LAZOGLU I., "Forces and hole quality in drilling", *International Journal of Machine Tools and Manufacture*, Vol. 45, 2005, p. 1271–1281.

[PIQ 00] PIQUET R., FERRET F., LACHAUD F., SWIDER P., "Experimental analysis of drilling damage in thin carbon/epoxy plate using special drills", *Composites. Part A: Applied Sciences and Manufacturing*, Vol. 31, no. 10, 2000, p. 1107-1115.

[RAH 99] RAHMAN M., RAMAKRISHNA S., PRAKASH J.R.S., TAN D.C.G. "Machinability study of carbon fiber reinforce composite", *Journal of Materials Processing Technology*, Vol. 89, 1999, p. 292-297.

[RAM 02] RAMKUMAR J., ARAVINDAN S., MALHOTRA K., KRISHNAMURTHY R., "An enhancement of the machining performance of GFRP by oscillatory assisted drilling", *International Journal of Advanced Manufacturing Technology*, Vol. 23, 2004, p. 240–244.

[SAN 88] SANTHANAKRISHNAN G., MALHOTRA S.K., KRISHNAMURTHY R., "Machinability characteristics of fibre reinforced plastics composites", *Journal of Mechanical Working Technology*, Vol. 17, 1988, p. 195-204.

[SAN 89] SANTHANAKRISHNAN G., MALHOTRA S.K., KRISHNAMURTHY R., "High speed steel tool wear studies in machining of glass-fibre-reinforced plastics", *Wear*, Vol. 132, 1989, p. 327-336.

[SCH 05] SCHMITT R., ORTH A., HAFNER P., "Analysis of current approaches for automatic measurements of reinforced composites", *Proceedings of the 38th CIRP International Seminar on Manufacturing Systems*, Florianopolis, 16–18 May 2005.

[SEI 07] SEIF M.A., KHASHABA U.A., ROJAS-OVIEDO R., "Measuring delamination in carbon/epoxy composites using a shadow moire laser based imaging technique", *Composite Structures*, Vol. 79, 2007, p. 113–118.

[SHA 84] SHAW M.C., *Metal Cutting Principles*, Oxford University Press, Oxford, 1984.

[SIN 06] SINGH I., BHATNAGAR N., "Drilling of uni-directional glass fiber plastic (UD-GFRP) composite laminates", *International Journal of Advanced Manufacturing Technolology*, Vol. 27, 2006, p. 870–876.

[SRE 99] SREEJITH P.S., KRISHNAMURTHY R., MALHOTRA S.K., NARAYANASAMY K., "Studies on the machining of carbon: phenolic ablative composites", *Journal of Materials Processing Technology*, Vol. 88, 1999, p. 43-50.

[STO 96] STONE R., KRISHNAMURTHY K., "Neural network thrust force controller to minimize delamination during drilling of graphite-epoxy laminates", *International Journal of Machine Tools and Manufacture*, Vol. 36, no. 9. 1996, p. 985-1003.

[TAG 90] TAGLIAFERRI V., CAPRINO G., DITERLIZZI A., "Effect of drilling parameters on the finish and mechanical-properties of GFRP composites", *International Journal of Machine Tools and Manufacture*, Vol. 30, no. 1, 1990, p. 77-84.

[TSA 03] TSAO C.C., HOCHENG H., "The effect of chisel length and associated pilot hole on delamination when drilling composite materials", *International Journal of Machine Tools and Manufacture*, Vol. 43, 2003, p. 1087–1092.

[TSA 04] TSAO C.C., HOCHENG H., "Taguchi analysis of delamination associated with various drill bits in drilling of composite material", *International Journal of Machine Tools and Manufacture*, Vol. 44, 2004, p. 1085-1090.

[TSA 07] TSAO C.C., HOCHENG H., "Effect of tool wear on delamination in drilling composite materials", *International Journal of Machine Tools and Manufacture*, Vol. 49, 2007, p. 983–988.

[TSA 08] TSAO C.C., HOCHENG H., "Evaluation of thrust force and surface roughness in drilling composite material using Taguchi analysis and neural network", *Journal of Materials Processing Technology*, Vol. 203, 2008, p. 342–348.

[VEL 05] VELAYUDHAM A., KRISHNAMURTHY R., SOUNDARAPANDIAN T., "Evaluation of drilling characteristics of high volume fraction fibre glass reinforced polymeric composite", *International Journal of Machine Tools and Manufacture*, Vol. 45, 2005, p. 399–406.

[WAN 95] WANG D.W., RAMULU M., AROLA D., "Orthogonal cutting mechanisms of graphite/epoxy composite. Part I: unidirectional laminate", *International Journal of Machine Tools and Manufacture*, Vol. 35, no. 12, 1995, p. 1623-1638.

[WIN 97] WINDHORST T., BLOUNT G., "Carbon-carbon composites: a summary of recent developments and applications", *Materials & Design*, Vol. 18, no. 1, 1997, p. 11-15.

[ZHA 01] ZHANG H., CHEN W., CHEN D., ZHANG L., "Assessment of the exit defects in carbon fibre-reinforced plastic plates caused by drilling", *Precision Machining of Advanced Materials*, Vol. 196, 2001, p. 43-52.

Chapter 4

Abrasive Water Jet Machining of Composites

This chapter is focused on abrasive water jet machining of composites. This chapter aims at defining the abrasive water jet technology through its history, the machining process, technologies, processes and applications.

4.1. Introduction

Abrasive water jet technology (AWJT) appears to be suitable for composite part machining, by completing the cutting tool capabilities [HAS 99]. Different research programs and many industrial observances have demonstrated the pertinence of the AWJ-composite association for cutting applications [WAN 99]. Currently, the main advantages are:

– low induced force on the material;

– low machining temperature;

Chapter written by François CÉNAC, Francis COLLOMBET, Michel DÉLÉRIS and Rédouane ZITOUNE.

168 Machining Composite Materials

– high cutting speed;

– few tooling;

– no risk of delamination for the right parameters (see section 1.6).

However, AWJT is a complex subject to approach because it deals with an hypersonic triple phase jet which does not impose a shape but a flow of energy.

4.2. Brief history of AWJT

Water jet technology was first developed in the 1950s, when a forestry engineer, Dr. Norman Franz, experimented with cutting lumber with an early form of water jet cutter. Following this, the main technological step occurred in the 1970s when Dr. Mohamed Hashish created a technique to add abrasives to the water jet cutter.

Nowadays, either a plain water jet or an abrasive water jet is used in every field of industry for cutting applications. Moreover, scientific investigations have demonstrated the capabilities of AWJT for milling [CEN 08], turning and piercing even if those techniques require industrial developments and applications.

4.3. AWJ machining process

AWJT (Figure 4.1) uses a high speed water jet (750 m/s) obtained when water under pressure (4,000 bar) passes through a small orifice (diameter: 0.33 mm). Grenat (80 mesh) is transported by air and introduced into the water jet (abrasive mass flow rate: 350 g/min). The particles are then accelerated by the water jet within a focusing tube (diameter: 1 mm) and thrown on the part.

Usually, only the feed rate is adjusted for a specific material, thickness and required quality. The impinging jet contains, in volume, 93% air, 6.5% water and less than 0.5% abrasive particles.

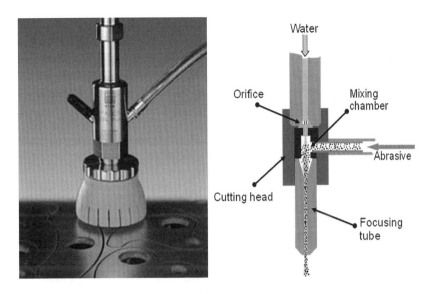

Figure 4.1. *Picture (left) and drawing (right) of an AWJ cutting head*

AWJT could be thought of as high speed local sanding. Indeed, only the particles are machining whereas the water is simply used to accelerate them through its initial momentum [4.1]:

$$\dot{m}_w \cdot v_w = \eta \cdot \rho_w \cdot \frac{\pi \cdot \phi_o^2}{4} \cdot \left(\frac{2 \cdot P}{\rho_w} \right) \qquad [4.1]$$

where $\dot{m}_w$ is the water mass flow rate, v_w is the water velocity, η is an efficiency coefficient, ρ_w is the water density, and ϕ_w is the orifice diameter.

At the exit of the focusing tube, this water initial power is partitioned between the fluid and the abrasive particles. Then, this abrasive water jet impacts the part and machines it through the following wear modes.

Water wedge mechanism

While impacting the material, the water acts as a wedge which propagates the upper-surface initial-micro-cracks until particles are removed (Figure 4.2). In order to be machined by a plain water jet, a part must contain micro-cracks and have lower micro-crack propagation minimum power than the water jet power.

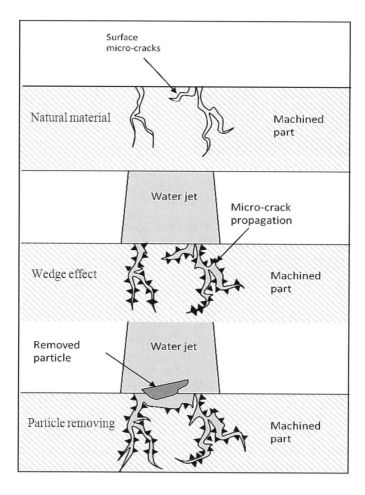

Figure 4.2. *Scheme of the water-wedge wear-process evolution with the different main phases (from top to bottom)*

Concerning the composite materials, the mode I delamination minimal power remains lower than the power required for propagating micro-cracks in the other directions. This implies that the water energy contained within the jet must remain lower than the delamination energy.

Abrasive wear mechanism

Finnie developed a wear model describing the erosion mechanisms by abrasive particles [FIN 58]. He described two wear modes (Figure 4.3): micro-cutting and impact lateral cracking. Micro-cutting mostly occurs for low impact angles and for ductile materials. The lateral cracking occurs for large impact angles and for brittle materials. Clearly, these mechanisms generally do not act separately, but in combination. Their importance for the particular erosion process depends on several factors, such as the impact angle, the particle kinetic energy, the particle shape, target-material properties and environmental conditions.

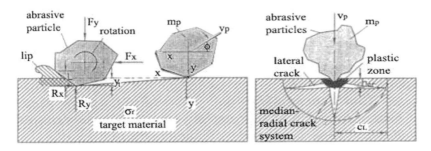

Figure 4.3. *Scheme of the abrasive wear processes; micro incision (left) and lateral cracking (right)*

4.4. AWJ cutting process

For workshop main applications, the cutting head is fixed to a moving system controlled through a computer. Depending on its quality, the water is simply 1 μm filtered plane water. Then, it is pushed under pressure using an intensifier to reach 400 MPa (600 MPa for the most recent applications). The water is then driven to the cutting head. After the cutting process, the jet is caught in a water tank (Figure 4.4). Then, the resulting mud is filtered to separate the water and the particles. In most cases, the particles are recycled for other applications (road works, etc.).

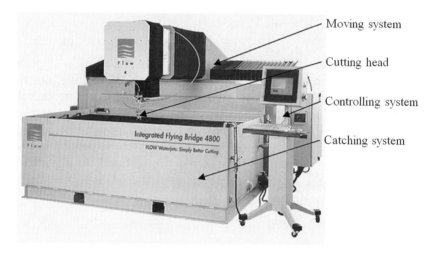

Figure 4.4. *Picture of an AWJ cutting machine (courtesy of FLOW Corp.)*

4.5. Quality of the kerf

Commonly, the main AWJ variable parameter is the feed rate. For a jet setting and a part (material and thickness), a maximal feed rate may be established. This is the idea of "workability": the trimming quality is a direct function of the set feed rate and the maximal feed rate ($\%f_{max}$). The main kerf quality criterion is the striation.

Figure 4.5 shows the striations for $\%f_{max} = 80\%$. At the top of the kerf, the jet mostly machines by micro-cutting. As the jet moves through the part, it loses energy, is deflected and a stair appears. Then, the stairs are pushed down by the jet, mostly using lateral cracking. Figure 4.5 shows a succession of the kerf fronts along the cutting process achieved on hard steel.

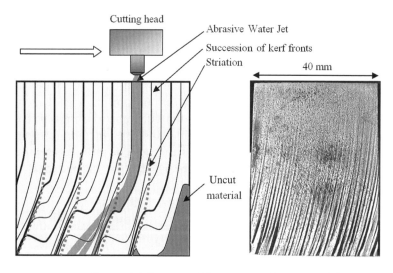

Figure 4.5. *Drawing of the striation phenomena (left) and picture of striations (right)*

4.6. AWJ cutting of composite materials

Composite materials can easily be trimmed using AWJ without the risk of delamination, moisture regain or pollution if the setting of the machine fits the application. The main current limit is the youngness of the technology. In fact, the knowledge and the knowhow required to machine composite materials with AWJ require a lot of experience to avoid the risks. Two AWJ aspects are generally critical for the cut-composite material-integrity: the piercing of the material and the striation phenomena. All those key points are linked to the water impinging energy. Currently, while cutting a part, it is the water jet local deflection that leads us to impose static pressure on the part and initiates wedge mechanism and delamination.

Piercing the material

The piercing of composite material is the main limit of the AWJ process for cutting composites. Every time it is possible to start cutting outside the part, the machining remains easier to perform. Currently, for inner cuttings, the part must be pierced by AWJ before it is cut. To do this, the first jet impact must contain abrasives to absorb the water power. This is allowed by vacuum systems which insert abrasives into the mixing chamber before water. The second problem with composite material AWJ drilling is that the jet has to turn back as long as the hole remains blind. Thus, the AWJ global power must remain low and the abrasive proportion high enough to keep the water impinging energy under the delamination limit. Table 4.1 presents the common cutting setting and the vacuum setting. Figure 4.6 shows the top of a composite part cut with a vacuum system (on the left) and without (on the right). The second part is delaminated during the piercing.

AWJ power main parameters	Composite cutting common setting	Composite drilling vacuum setting
Pressure	4,000 bar	1,000 bar
Orifice diameter	0.33 mm	0.25 mm
Abrasive mass proportion	11% (350 g/min)	16% (150 g/min)

Table 4.1. *Common AWJ cutting and vacuum settings for piercing composites*

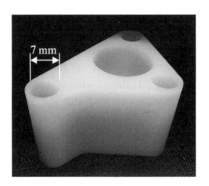 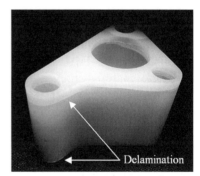

Figure 4.6. *Composite part cut using a vacuum system (left) and the same part cut without a vacuum system (right)*

While using a vacuum system, local delamination of the peel ply may occur at the exit of the piercing and penalizes the visual aspect. This is due to a very low mode I delamination energy of this ply. This phenomenon is shown Figure 4.7. It is not critical for the part mechanical behavior, but it has a negative affect on the appearance.

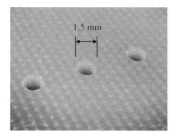
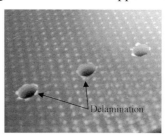

Figure 4.7. *Top face of a pierced composite part (left) and bottom face (right)*

Striation phenomena

The cutting %f_{max} must be kept low enough to avoid large striations. Currently, the striations are linked with a deflection of the jet which may lead to delamination if the abrasive proportion is not high enough for any reason (pressure too high, orifice too large and/or not enough abrasive). Figure 4.8 shows the kerf obtained with the same jet power and constitution, but with different feed rates on a glass/epoxy composite. It appears that if the %f_{max} is too low, the deflection of the jet is high and delamination occurs at the bottom of the kerf.

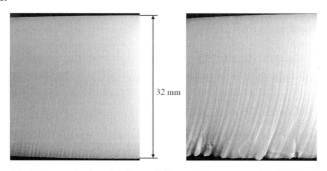

Figure 4.8. *Picture of a low feed rate (%f_{max} = 40%) cut profile (left) and picture of a high feed rate (%f_{max} = 80%) cut profile with striation and delamination on the bottom (right)*

4.7. Applications

AWJT is currently mostly used for the following reasons:

– the tool thickness is more or less 1 mm; for expensive materials, it may be possible to save more material or to do more parts on one only plate;

– multi-materials may be machined, as well as homogenous materials. There is no optimal cutting speed related to the material. If the materials are very different, as regards the AWJ workability, the quality may be decreased;

– the induced force on the material remains very low, depending on $\%f_{max}$. In the worst situation ($\%f_{max}$ = 100%), around 50 kg is imposed on the axial direction. With a 4,000 bar AWJ machine, the machined plate does not have to be kept fixedly in position while the jet moves;

– the machining temperature remains low; when machining very hard and thick materials, the temperature may reach 100°C;

– the cutting-head wear does not depend on the machined material but only on the machining time;

– the cutting speed is high: around 5 m/min for a 5 mm-thick carbon-epoxy composite to obtain a good quality.

However, on the other hand the AWJT currently has limits which penalize its development:

– the AWJ is mostly limited to 2D trimming. 5-axis machines exist, using a catcher to bring back the jet after it passes through the part. However, this promising technology and the associated processes are still in progress;

– the machining tolerance may reach +/-0.05 with a good roughness (< 2 µm) with precautions, but it is very difficult to do better;

– the AWJT uses water. This may be a problem, depending on the material and/or application;

– the abrasive particles may pollute the part. There is no inclusion due to impact of abrasive within the machine part. However, porous materials may have to be cleaned up if necessary;

– the visual appearance shows the use of AWJT. Machining composite materials using AWJ is not widely accepted yet;

– AWJT is still unknown and involves new manufacturing processes that are difficult to integrate.

Figure 4.9 shows pictures of AWJ divided into composite parts.

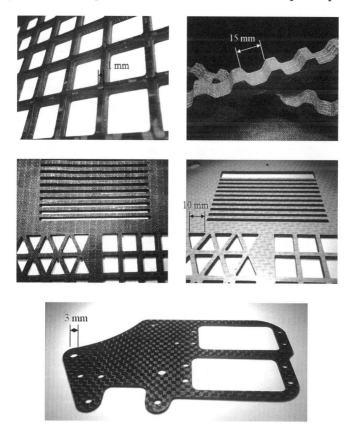

Figure 4.9. *Pictures of AWJ cut organic composite and CMC (top right) parts achieved in JEDO Technologies (Labège, France)*

4.8. Perspectives

The main forthcoming developments in AWJT concern the 5-axis machines. Indeed, this type of machine is already used, but it requires much more development to be able to exploit all the technical capabilities of 3D AWJ machining.

Another current evolution concerns the reliability of the technology. The current machines require very vigilant operators in order to assure that no trouble occurs. The development of control tools is already helpful.

AWJT is finally being developed by finding new applications such as micro-piercing, turning and milling.

For each of these processes, application domains are being discovered for all materials in order to identify the added values and real limits of AWJT.

4.9. AWJ milling of composite materials

The capabilities of AWJ technology are widely used for composite material cutting. Nevertheless, it is almost never used for blind machining although it appears to be even more interesting.

The principle (Figure 4.10) is to reduce the impinging energy by both decreasing the water power (through the water pressure or the orifice diameter) and decreasing the exposure time (through the feed rate). This leads to an incision.

Then this incision may be enlarged by increasing the standoff distance. At the end, the part is scanned to achieve milling. In order to obtain a good geometry, the milled area may be delimited by a mask.

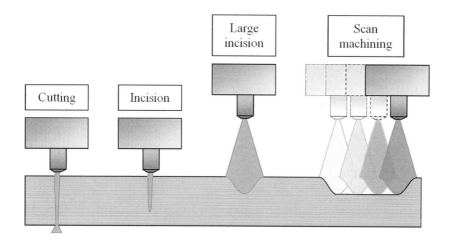

Figure 4.10. *Drawing of the AWJ milling principle*

The main interests of AWJ milling of composites are:

– the AWJ machines a depth from the upper impinged surface whereas conventional processes machine a dimension from a static reference;

– the standoff distance is slightly influenced by the machined depth so that a bent part may be machined with a two-axis AWJ machine-tool;

– as long as the jet energy distribution between the particles and the water is mastered, there is no risk of delamination, pollution and/or moisture regain.

This process is mainly limited by:

– the limited number of studies;

– the current technology that is not completely adapted for milling applications.

Nevertheless, AWJ milling (Figure 4.11) of composites opens up a new field of investigation because it allows ply per ply machining of bent composite parts (HexPly M21T700) and may be very convenient to achieve stepped lap repair machining (achieved at JEDO Technologies, Labège, France).

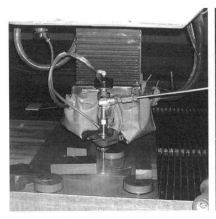 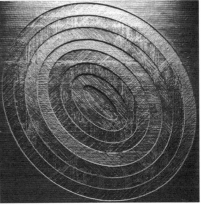

Figure 4.11. *Picture of AWJ milling of composite (left) and picture of an AWJ M21T700 8-ply stepped lap repair machining (right)*

4.10. References

[CEN 08] CÉNAC F., COLLOMBET F., ZITOUNE R., DELERIS M., "Abrasive-water-jet blind-machining of polymer matrix composite materials", *Proceedings of ECCM13*, Stockholm, Sweden, paper 1728, 2008.

[FIN 58] FINNIE I., "The mechanism of erosion of ductile metals", *Journal of Engineering Material and Technology*, p. 527-532, 1958.

[HAS 99] HASHISH M., "Status and potential of waterjet machining of composites", *Proceedings of 10th American Waterjet Conference*, Huston, Texas, paper 64, 1999.

[WAN 99] WANG J., "Abrasive waterjet machining of polymer matrix composites – cutting performance, erosive process and predictive models", *International Journal of Advanced Machining Technology*, Vol. 15, p. 757-768, 1999.

Chapter 5

Machining Metal Matrix Composites

5.1. Introduction

Metal matrix composites (MMCs) have high performance [PRA 06], and hence have the potential to replace conventional materials in many engineering applications [DAV 07] such as in automotive and aerospace structures [PRA 07]. Very often, the metal matrix materials of MMCs are aluminum alloys, zinc alloys, copper alloys and magnesium alloys, while the reinforcement materials are silicon carbide (SiC), aluminum oxide (Al_2O_3), silimullite, glass fibers, graphite, fly ash and talc in the forms of whiskers, fibers or particulates [PRA 08c, BAS 06]. MMCs, particularly aluminum-based particle/fiber-reinforced composites, have a high strength to weight ratio and wear resistance [ZHA 95a, ZHA 95b, ZHA 95c, YAN 95, ZHA 97], thus most of the investigations are on SiC or Al_2O_3 particle-reinforced aluminum matrix composites.

There is little in the literature related to machining of MMCs before 1990 [HEA 01]. Extensive research on machining of aluminum alloy MMCs started in the 1990s. While the reinforced particles bring

---

Chapter written by Alokesh PRAMANIK and Liangchi ZHANG.

MMCs superior physical properties to MMCs, they also cause very high tool wear and inferior surface finish in a machined surface [PRA 07, PRA 08]. Although the application of near net shape forming and modified casting can reduce the requirement of machining operations, they cannot be eliminated [BAS 06]. Therefore, difficulties associated with the precision and efficient machining of MMCs have become an important issue [PRA 06]. This chapter will discuss some state-of-the-art developments in understanding the material removal mechanisms of MMC machining.

5.2. Conventional machining

Conventional machining methods such as turning, drilling, grinding and milling have been applied to composite materials using different tools and cutting conditions.

5.2.1. Turning

Most of the research related to machining of MMC can be attributed to turning because it is a most common and simple process. The principal machining parameters that influence the turning quality are cutting speed, feed rate, depth of cut, reinforcement properties and geometry and material of the cutting tools. Generally, polycrystalline diamond (PCD), cubic boron nitride (CBN), TiC, Si_3N_4, Al_2O_3, and WC are used as the cutting tool materials.

Tool wear during turning MMC is generally high due to the presence of hard reinforced particles. Tool pitting, chipping, microcracking and fatigue [ELG 00, DER 01] are the most commonly observed phenomena. It has been argued that the dominant wear mechanisms during turning of MMC are two-body and three-body abrasions and these are due to hard reinforced particles and debonded tool material grains [HEA 01, DAV 02, CHA 96, DAV 00, DIN 05, YAN 00, HOO 99, WEI 93]. These abrasion processes result in wear in primary and secondary flanks [XIA 01, DAV 00, WEI 93]. Some researchers [HOO 99, AND 00] have noted that adhesion is also a cause of tool wear, as thin films of the workpiece material have been found to adhere to the worn areas. Chemical wear during machining of

MMCs has not been reported for any tool material. This is because the constituents of MMCs, i.e. the matrix material and reinforced particles (e.g., SiC and Al_2O_3) are chemically inert to almost all cutting tools (e.g., PCD, CVD-coated, carbide tools) under the cutting conditions.

Among all the cutting tools, the coarse-grained PCD inserts perform better in terms of tool wear and surface roughness of machined workpieces [CHA 96]. Cutting speed is one of the important factors that influence the tool wear. At lower cutting speeds there is a strong tendency to form a stable built-up-edge (BUE) on a tool surface [MAN 03, 03b]. Tool wear increases with the increase of cutting speed, particulate size and volume percentage of reinforcements [CIF 04, JOS 99, LOO 92, XIA 01, CHA 96].

Generally, tool wear increases with the increase of feed [LIN 95], though some researchers noted reduced tool wear at higher feeds [TOM 92]. The rate of increase of tool wear with feed is highly dependent on the cutting speed. It is noted that at a low speed the flank wear increases marginally with feed, but at a higher speed tremendous increase of flank wear is noted with the increase of feed [MAN 03, DAV 02].

Some models are available in the literature where the influence of the particles was considered explicitly for predicting tool wear. For example, Pedersen *et al.* [PED 05] proposed an empirical wear model to predict flank wear based on the probability of four possible tool–particle interactions depending on the tool–particle orientations. Kishawy *et al.* [KIS 05] developed a model to predict flank wear rate considering two-body and three-body abrasion between the MMC and tool. Kannan *et al.* [KAN 05] proposed a tool wear model to predict flank wear by correlating hardness and energy consumption. Though these models show good agreement with investigators' experimental results, a comprehensive comparison with experimental results available in the literature is yet to be carried out [PRA 08c].

At a higher feed, surface roughness of a machined workpiece is controlled by feed rate and particle size, but at a low feed it is controlled only by the size of the reinforcement particles in an MMC

[PRA 08b]. There are contradictory reports on the effect of speed on the surface roughness of MMC. Researchers have noted a different trend in surface roughness with the variation of speed [BAS 06]; but the fact is that the variation of surface roughness with the speed is not significant and it is comparable to the size of reinforcements at a lower feed [DAV 02, LIN 95, MUT 08]. The correlation between speed and roughness is mainly dominated by feed and uncontrollable interactions of reinforcements with the cutting tool and newly generated surface [PRA 08, SAH 02].

In most cases, the cutting speed does not influence cutting forces [PRA 08]. There are only a small number of cases where an increase in cutting speed resulted in a slight decrease in the cutting forces, but the evidence is not solid [WAN 03]. The feed has a significant influence on machining forces, as the forces rise remarkably with the increase of feed [PRA 06, PRA 08]. Several force prediction models have been developed for cutting MMCs. For instance, Kishawy *et al.* [KIS 04] developed an energy-based analytical model to predict the forces in orthogonal cutting of an MMC using a ceramic tool at a low cutting speed. Pramanik *et al.* [PRA 06, PRA 08d] established a mechanics model for predicting the forces of cutting ceramic particle-reinforced MMCs based on the force generation mechanisms of chip formation, matrix ploughing and particle fracture/displacement.

Similar to monolithic materials, the increase of depth of cut increases tool wear, machining forces and surface roughness during machining of MMCs [MAN 03, DAV 02, PRA 06]. This is due to the increase of material removal with the increase of depth of cut.

5.2.2. *Drilling*

Due to the poor machining properties of MMCs, drilling is a challenging task [RAM 02]. Similar to other conventional machining processes, many problems are associated with the drilling of MMCs, such as tool wear, high drilling forces and burr formation [MON 92, MUB 95, MUB 94]. Tool geometry and materials, reinforcement properties and cutting conditions affect the drilling outcomes. High-speed steel (HSS), carbide, cubic boron nitride (CBN) coated and polycrystalline diamond (PCD) tools are generally used for drilling MMCs [RAM 02, MUB 95].

The performance of HSS tools in drilling MMCs is generally poor. Very high wear rate of this type of tool makes it unsuitable for MMCs. This is due to the fact that the hardness of reinforced ceramic particles is much higher than that of HSS [DAV 01]. Ramulu *et al.* [RAM 02] noted 0.84 mm flank wear and could drill only one hole on 75 mm thick 20 vol.% $(Al_2O_3)p/Al6061$ at 1,320 rpm and 0.0635 mm/rev but several holes were possible on 10 vol.% $(Al_2O_3)p/6061$. A lower volume percentage of reinforcements causes less interaction between tool and particles, hence higher performance of an HSS tool. The wear rate increases with the decrease of feed. This could be due to the longer contact time between the drill edge and machined surface [RAM 02]. It has been reported that during drilling of MMCs, cutting tools experience four types of problems such as abrasion, chipping, groove formation and build-up-edge.

The maximum wear takes place at the outer edges of the tip and minimum wear occurs at or near the drill tip. The maximum rotational force and the maximum contact with the workpiece occur away from the drill tip and thus the outer edge is abraded more quickly [MUB 95]. Flank wear is common to all types of drilling tools.

In addition to flank wear, margin wear occurs in softer tools and tools with low clearance angle. This is due to the rubbing action between the drill tip sides and the hole surface. This produces bad surface finish [RAM 02, MUB 95]. Due to marginal wear the exit side of the hole is not cut cleanly but forced out to form a burr. Micro-chipping is mainly noted at the outer corner of a drill tool. Edge chipping occurs due to interaction of the hard and abrasive reinforced particles in an MMC with the tool edges [DAV 01]. BUE normally generates on chisel edges and flank edges [RAM 02, MUB 95]. Many uniform grooves are formed on the flank face due to ploughing by hard reinforcement while chips flow over the rake face. Due to the superior hardness of PCD drills, their resistance wear is much higher [RAM 02, MUB 95, HEA 01].

Mubaraki *et al.* [MUB 95] quantitatively compared the performance of HSS, carbide and PCD tools under similar cutting conditions (speed 1,800 rpm, feed rate 110 mm/min, depth of drill 20 mm). They found that for the HSS drill a flank wear of 1.00 mm is reached in drilling for as little as 12 s. For the WC drill, a flank wear of 0.16 mm was observed after drilling for a period of 600 s and in the case of the PCD drill, after 2,210 s of drilling, the flank wear was only 0.12 mm.

Drilling forces depend on the work materials, tool materials and cutting conditions. With the increase of the reinforcement volume fraction, the drilling forces increase significantly. Regardless of the tool material and reinforcement volume fraction, thrust force, torques and surface roughness are highly dependent on feed rate while cutting speed generally influences the drilling forces insignificantly [RAM 02, TOS 04, MOR 95, DAV 01]. A better surface finish and a lower drilling force are generally achievable by a harder tool. The application of a softer tool is often associated with a high wear that blunts the tool's cutting edge, resulting in a comparatively higher drilling force and a worse surface finish [TOS 04]. With the increase of the point angle of the drill tool, the surface roughness decreases [TOS 04, TOS 04a].

Davim *et al.* [DAV 03] established some empirical equations using a multiple linear regression analysis to calculate tool wear, specific cutting pressure and hole surface roughness in terms of cutting velocity, feed rate and cutting time, i.e.

$$VB = 0.029 + 0.163f + 0.667 \times 10^{-3}V + 4.000 \times 10^{-3}T, R=0.83$$

$$K_s = 6591.4 - 17013.7f - 60.3V + 129.6T, R=0.86$$

$$R_a = 0.587 + 2.588f - 1.911 \times 10^{-3}V + 8.130 \times 10^{-3}T, R=0.82$$

where VB is the tool wear (mm), K_s is the specific cutting pressure (N/mm^2), R_a is the arithmetic mean roughness (μm), f is the feed (mm/rev), V is the cutting velocity (m/min) and T is the cutting time (min). They used the MMCs of A356/20/SiCp-T6 (aluminum with 7.0% silicon, 0.4% magnesium, reinforced with 20 vol.% particles of silicon carbide (SiC)) heat treatment (solutionizing and ageing T6-5 h at 154°C). The average dimension of the SiC particles was about 20 μm. PCD tools were used to cut holes (diameter of 5 mm) in 15 mm thick MMC discs.

5.2.3. *Grinding*

Grinding is usually the subsequent machining process to achieve the desired geometry, assembling tolerance and surface integrity of a near-net shape component. It is particularly needed to acquire high dimensional accuracy and surface finish. The grinding of MMCs, however, has received little attention in research. Comprehensive information on grinding of MMCs is not available. In studying the grinding of MMCs, researchers have used different materials and conditions. As a result, it is difficult to make reasonable comparisons and conclusions.

Grinding does not perform very well on soft materials because of the tendency of the chip loading to wheels [ILI 96]. However, due to the improved chip breakability of MMCs, grinding of MMCs can be done reasonably [ILI 96, PRA 08b]. Different types of grinding wheel materials have been tested on MMCs, such as silicon carbide, alumina, CBN, diamond (resin bonded and electroplated), etc. [ILI 96, CHA 99, ZHO 02, RON 09].

Generally, lamella structured curled chips with no hollow sphere are generated during the grinding of MMCs composed of 15, 20 or 25 vol.% particulate and 20 vol.% whisker-reinforced aluminum (Al 2009) [ILI 96]. The grindability is better compared with that of a non-reinforced aluminum alloy in terms of a better surface finish and a lower tendency of chip loading to wheel [ILI 96]. Super abrasive wheels of appropriate grit and binder are often desirable because chip loading and attrition wear would be lower [RON 09]. Binding materials of grinding wheels play an important role in the self-sharpening process of wheels. If diamond abrasives are soft bonded, the self sharpening (partial fracture, dislodging) can be performed easily, but it is expensive. On the other hand, self sharpening is not easy to carry out for hard bonded diamond abrasives, resulting in degradation of the grinding wheel and worse performance. Grinding wheels bonded by hard bonding, such as electroplated wheels, experience a relatively higher grinding force, acoustic emission energy and surface roughness compare to that with softer bond materials (e.g. resin) [RON 09].

As the percentage of the reinforcement increases, the hardness of MMCs increases and the grinding wheel degrades quickly. The grinding forces show an increasing trend with the wheel degradation and hardness [ZHO 03]. Grinding forces generally increase with the increase of depth of cut [ILI 96, KWA 08]. Workpiece speed has negligible influence on surface roughness and grinding forces.

The material properties such as hardness and type of reinforcement, i.e. particles or whiskers, influence the surface texture of the ground surfaces. The volume fraction of the reinforcement as well as the shape and dimension of the reinforcement material, play an important role in both surface texture and tool wear [ILI 96]. Smearing of the aluminum matrix on the ground surfaces has been noted during rough grinding, but it reduces during fine grinding [ZHO 03, HUN 97, ZHO 00].

Grinding imposes compressive stresses in both constituents and creates a macroscopic compressive zone in the near surface region. As the depth increases, the effect of grinding diminishes, and the residual stresses in both constituents gradually return to the annealed levels [LEE 95]. Work hardening of matrix material is generally limited to the depth equal to the diameter of reinforcements approximately [HUN 97]. Diamond wheels generate lower forces and surface cracks compared to other abrasive materials. In composite materials the spread of the cracks are arrested by the presence of reinforced particles [CHA 99].

According to Zhong *et al.* [ZHO 03, ZHO 02], SiC wheels can be used for rough grinding of alumina/aluminum composites, because SiC is harder than Al_2O_3 and much less expensive than diamond. Rough grinding with a SiC wheel followed by fine grinding with a fine-grit diamond wheel is recommended for the grinding of alumina/aluminum composites.

5.2.4. Milling

Similar to the grinding of MMCs, little information is available on the milling of MMCs. Polycrystalline diamond, carbide and coated (TiAlN, TiN + TiAlN, TiCN + Al_2O_3 + TiN) carbide tools have been reported to mill MMCs [RED 08, KAR 06, ÜBE 07, ÜBE 08]. Higher tool wear is generally noted for milling MMCs, due to the hard reinforcement particles. Tool wear reduces when harder tool materials are used. Enhanced machinability of MMCs is noted compared to a non-reinforced aluminum alloy during milling [RED 08]. Though the tendency to clog the cutting tool is very low, workpiece material adhesion appears a little distance away from the tool tip along the rake and clearance face on the tool tip [RED 08].

Under similar machining conditions, MMCs (Al/SiC) give better surface finish than that of matrix materials. Machined surface roughness and forces increase with the increase in feed. Surface roughness decreases with the increase of cutting speed. Compressive residual stresses are generated in the surface of milled MMCs. It is seen that tool wear increases almost linearly with the increase of chip volume. At a low speed, feed does not influence flank wear significantly but at higher speeds flank wear decreases with the increase of feed [ÜBE 07].

Abrasion and adhesion are the main wear mechanisms for all of the tools mentioned above. Similar to the turning operation, BUE formation is observed at low cutting speeds but its extent decreases with the increase of the speed.

During the milling of MMCs, coated carbide tools perform better than uncoated ones at all cutting speeds and feed rates tested. Performance of carbide milling tools increases with the increase of coating layer and thickness. Tool wear increases with the increase of cutting speeds and decreases with the increase of feed. Generally, low cutting speeds and high feed rates are desirable to reduce tool wear during milling aluminum MMCs [ÜBE 07, ÜBE 08, KAR 06].

5.3. Non-conventional machining

Higher tool wear and worse surface finish in conventional machining significantly hinder the use of MMCs. Electronics-grade MMCs of high reinforcement content are nearly impossible to machine by conventional methods. Thus, non-conventional techniques, including electro-discharge, laser-beam, electro-chemical and water jet machining have also been applied to these materials [HIH 03].

5.3.1. *Electro-discharge machining*

Electro-discharge machining (EDM) is a multipurpose process for machining intricate or complex shapes in conducting materials where material is removed by erosion caused by electrical discharge between the electrode and the workpiece [ABR 92]. The EDM process takes place in a dielectric fluid where the tool is one electrode in the shape of the cavity to be produced and the workpiece to be machined is the other electrode. The tool is then fed toward the workpiece in a controlled path to produce the shape of the electrode or its movement. In wire electro-discharge machining (WEDM) a thin wire is used as the tool electrode.

Generally, the machining characteristics of WEDM of MMCs are similar to those occurring in a matrix material but slower. This is because of the decrease in thermal and electrical conductivity of MMCs caused by the non-conductive reinforcements.

During WEDM the feed of composites significantly depends on the type and amount of reinforcement. The maximum cutting speed achievable on MMCs reinforced by SiC and Al_2O_3 particles are approximately 3 times and 6.5 times lower than that on the matrix material [ROZ 01]. The machining time for materials with 25% fiber reinforcement is almost double than for those with 15% fibers [RAM 89]. Material removal rate is generally higher at the beginning of machining but slows down due to the entrapment of reinforcement particles in the spark gap [HOC 97]. Thus, side sparking is induced and this deteriorates the dimensional stability. Feed rate, material removal rate, surface roughness and tool wear increase with the increase of current, voltage and pulse ON-time [ROZ 01, SIN 04, RAM 89, MOH 04].

At a low velocity of dielectric fluid, short-circuiting becomes less pronounced due to the accumulation of particles into the spark gap. This improves the material removal rate. On the other hand, a higher velocity/flushing pressure obstructs the formation of ionized bridges into the spark gap. This causes a higher ignition delay and decreased discharge energy, thus reducing the material removal rate. Tool wear reduces at higher velocity/flushing pressure because of the higher cooling rate of the tool [SIN 04]. In the case of electro-discharge drilling (using hollow tool), the material removal rate, tool wear rate, surface roughness and cutting feed rate increase with the increase of flushing pressure [YAN 99, WAN 00]. Although eccentric through-hole in a rotational electrode during drilling blind hole gives a higher material removal rate, the tool wear rate is higher [WAN 00]. For effective EDM of MMCs, large current and short on-time has been recommended [HOC 97].

The matrix materials surrounding the reinforcement melt during EDM. The resolidified machined surface, usually known as the recast layer, generally contains a large number of micro-cracks on the surface. Random voids are also seen at the recast layer which may be due to the imperfect joining of the molten droplets or trapped gas during solidification. Surface damage is considered to be limited to the thickness of the recast layer [HUN 94]. Reinforcement particles are the least present on the recast layer, as most of the particles may get raked-up and removed during machining. At higher voltage and current, it is more pronounced and hence the reinforced particles may be deposited below the recast layer [SIN 04].

An electrode with higher melting temperature has better wear resistance during EDM. For example, the wear rate of brass is greater than copper due to the higher melting temperature of copper [RAM 89]. The tool wear rate on the negatively connected electrode is lower than the positively connected electrode irrespective of the electrode material and the volume percentage of reinforcements. A rotary electrode increases the material removal rate, decreases electrode wear and improves surface finish compared to a stationary electrode [MOH 02, MOH 04].

5.3.2. Laser-beam machining

A laser beam is often used for the machining of metals, ceramics and composites for faster processing and has the ability to obtain complex shapes. This method provides high rates of heating from a highly controlled source of energy which melts and vaporizes the workpiece material. Very few reports are found in the literature related to laser-beam machining of metal matrix composites. Composites based on a high thermal conductive matrix are characterized by a low absorption factor and are generally machined by a CO_2 laser beam [KAG 89, DAH 89, HON 97].

At a given power level of a laser beam, higher speeds mean that a lower amount of energy is available to remove the material in the cut zone. Thus, the depth and width of the kerf decrease with an increase in the cutting speed. At a higher cutting speed the machined surface is very rough and straddled with striations due to the unsteady motion of the molten layer or intermittent plasma blockage.

High volume SiC content 6061 Al MMC can be successfully cut by a laser to achieve a smooth cutting surface and a narrow heat-affected zone at moderate cutting speeds with an argon shielding gas [HON 97]. For instance, three distinct regions are produced in the heat-affected zone when cutting SiC-reinforced MMCs. In this case, plate/needle-like phases with small SiC particles and blocky Si particles were found close to the cutting surface with a narrow width of 50-60 μm.

Next to this region (70 μm in width) SiC particles were found to have redistributed with increased size and smoothed edges. Some large blocky Si, fine cellular/dendritic Al structures and Al/Si eutectics are apparent in this region. This region was then followed by plate/needle-like phases (8-10 μm in size) nucleated at the surface of the SiC particles. Unmelted base materials were present next to this region. The intense heat during laser machining melted the constituent materials and caused chemical reactions among them. The chemical reaction between SiC and Al produces the plate/needle-like phases (Al_4C_3, Al_4SiC_4) and free Si. Al_4SiC_4 mainly appeared in the form of large platelets and its growth proceeded by solid state diffusion [HON 97, DAH 89]. The extent of reaction between reinforcement and matrix can be controlled by the laser energy input [DAH 89].

5.3.3. *Electro-chemical machining*

In electrochemical machining, the material is removed by controlled electrochemical dissolution process of a workpiece. A voltage is applied while keeping the workpiece and tool at anodic and cathodic ends respectively in an electrolyte solution. During this process, the tool is advanced towards the workpiece in a defined path and the required shape of workpiece is achieved by using a suitable tool. Calomel and aqueous sodium nitrate solution have been used as cathode and electrolyte respectively. Dissolution occurred by the electrolytic removal of the matrix material, while the inert reinforced particles are flushed away by the electrolyte [HIH 03]. There are several attractive advantages of electrochemical machining such as no burrs, no stress, a longer tool life, damage-free machined surface, etc. [SEN 09].

During electrochemical machining, the material removal rate increases with the increase of the applied voltage, feed, electrolyte concentration and flow rate. Increased voltage and electrolyte concentration result in a higher machining current in the inter-electrode gap. Increased feed also increases the current density due to the reduction in inter-electrode gap. At a higher electrolyte flow rate, ions from the metal to the solution are more mobile to speed up the chemical reactions, thus the metal removal rate increases [SEN 09].

Unsteady and non-uniform metal dissolution leads to a poor surface finish at a low voltage and feed. At a higher feed, however, pit formation takes place due to the higher current densities and the presence of particles [HIH 03]. Excessive heating causes deterioration of the workpiece surface at a voltage above a certain limit. At a lower electrolyte flow rate, the ions of the material move slowly and produce streaks on the surface. Due to the depletion of ions, a poor surface finish is generated at a low electrolyte concentration [SEN 09].

The reinforcements do not significantly affect the breakdown potential of matrix materials. Hydrogen bubbles are produced during electro-chemical machining which impede dissolution and cause a nodular surface profile. The nodular profile can be eliminated by introducing rotational speeds at cathode and high electrolyte convection rates to flush away hydrogen bubbles. A very precise material removal can be made by accurately controlling the dissolution current and feed [HIH 03].

5.3.4. *Abrasive water jet machining*

The impact of solid particles is the basic event in material removal by abrasive water jet machining where a jet of high pressure and velocity, water and abrasive slurry is used to cut the target material by means of erosion [SHA 02]. Abrasive water jet (AWJ) technology has received considerable attention in the machining of difficult-to-machine and thermally sensitive materials. Machining with an AWJ has some advantages. In comparison to thermal machining processes (laser, EDM), AWJ does not induce high temperatures and as a consequence there is no heat-affected zone [MÜL 00]. It is thought to be a very fast machining process for MMCs as high feed rates are possible in this process [HAS 95]. However, it is very difficult to produce a workpiece with high geometrical accuracy using AWJ.

Generally, a rough surface is generated from AWJ machining [HAS 95, CAP 96, MÜL 00]. A smoother surface can be obtained with lower feed rates and depends on the size of abrasive particles used. Striation formation due to cutting lag and step removal is generally present when machining thicker MMC samples. The material removal process occurs mainly by cutting. The ductile shearing of the matrix material is observed from the abrasive scooping and ploughing path. Reinforcement particles in an MMC workpiece cut by AWJ are often pulled out from the matrix material if these are smaller than the abrasive particles [MÜL 00].

5.4. Tool–workpiece interaction

In most mechanical machining processes introduced in the previous sections, materials are removed through a cutting edge by externally applied machining forces. Thus, an understanding of the deformation of MMCs at the tool–workpiece interaction zone will be useful. From the previous sections it is also clear that tool–particle–matrix interactions have significant influence on the machining performance of MMCs such as tool wear, force generation and surface roughness.

Since analytical and/or experimental methods are not able to handle this type of investigation, the finite element method was used by Pramanik *et al.* [PRA 07] to explain the tool–particle–matrix interactions during machining of particulate reinforced MMCs. In doing so, they categorized the interaction between the tool and reinforcements into three scenarios: particles along the cutting path, particles above the cutting path and particles below the cutting path (Figure 5.1).

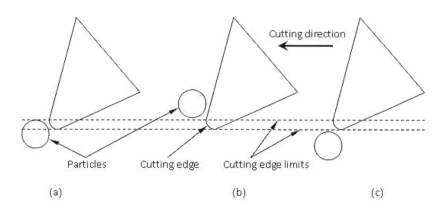

Figure 5.1. *Particle locations with respect to the cutting path: particles (a) along, (b) above and (c) below the cutting path [PRA 07]*

5.4.1. *Evolution of stress field*

When a particle is along the cutting path and interacts with the lower part of the cutting edge (i.e., the center of particle is below the center of the cutting edge), the compressive and tensile stresses are perpendicular and parallel respectively to the cutting edge in the matrix and particle in front of the cutting edge at the start of machining (Figure 5.2). This type of stress distribution may initiate fracture in the particle and debonding at the interface.

With the advancement of the tool, the matrix between the upper part of the particle and tool becomes highly compressive while the lower right interface of the particle becomes highly tensile (Figure 5.2(a)). This indicates that an anticlockwise moment is acting on the particle, thus debonding of the particle may be expected with further advancement of the tool.

When tool–particle interaction occurs, significant tensile and compressive stresses that are perpendicular to each other are found in the left part of the particle (Figure 5.2(b)). However, the right part of the particle is only under compressive stress. Such stress distribution may initiate particle fracture if the stresses are high enough. With further advancement of the tool, the particle debonds and ploughs through the matrix, making a cavity, then slides on the cutting edge and flank face (Figure 5.2(c)) and becomes almost stress-free (Figure 5.2(d)).

The particle located at the upper part of the cutting edge moves slightly upwards (Figure 5.2(c) & (d)) due to the plastic flow of the matrix. Initially the matrix in between the particle and tool is under highly compressive stress acting parallel to the cutting direction with no tensile stress (Figure 5.2(c)). On the other hand, parts of the particle and interface are under compressive stress along the cutting direction and under tensile stress perpendicular to the cutting direction.

This type of stress distribution can lead to particle debonding and/or fracture. After interacting with the tool's rake face, the particle partially debonds and moves up with the chip. With further advancement of the tool (Figure 5.2(d)) it then interacts with a nearby particle and consequently both particles are under highly compressive stress applied perpendicularly to the rake face. This highly compressive stress may cause fracture of the particle as well as wear on the tool rake face.

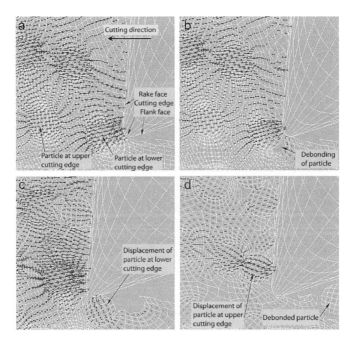

Figure 5.2. *Evolution of stress fields for particle along the cutting path. Compressive and tensile stresses are represented by black >–< and white <—> symbols, respectively. Their lengths represent comparative magnitudes [PRA 07]*

For particles above the cutting path, a highly compressive stress field perpendicular to the tool rake face through the particle and in the matrix in between particle and rake face (Figure 5.3(a) & (b)) is noted at the start of machining. At the same time, part of the particle and interface are under compressive (perpendicular to rake) and tensile (parallel to rake) stresses as shown in Figure 5.3(a). As stated before, this type of stress distribution may initiate particle fracture and interface debonding. As the tool proceeds, it interacts and partially debonds the particle. The contact region with the rake face is under highly compressive stress, hence fracture of the particle can be expected. At this stage the matrix in between this particle and the next one is also under very high compressive stress. With further advancement of the tool, the first particle interacts with the next particle and moves up along the rake face under highly compressive stress (Figure 5.3(b)).

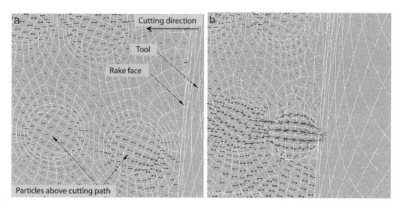

Figure 5.3. *Evolution of stress fields for particle above the cutting path. Compressive and tensile stresses are represented by black >–< and white <—> symbols, respectively. Their lengths represent comparative magnitudes [PRA 07]*

The stress distribution in the particle and matrix below the cutting edge has a direct influence on the residual stress of the machined surface. As the tool approaches the particle, the matrix in between the cutting edge and particle is under compressive stress acting in a radial direction to the cutting edge (Figure 5.4(a-c)). However, the particle and particle matrix interface are under compressive and tensile stresses which are acting in a radial direction to the cutting edge and parallel to the cutting edge respectively (Figure 5.4(a)). While the tool is passing over the particle, the direction of compressive stress remains radial to the cutting edge.

On the other hand, the direction of tensile stresses in the particle becomes parallel to the machined surface (Figure 5.4(b)). At the same time, the magnitudes of both stresses have decreased. It is also noted that the newly generated surface is under compressive residual stress which is parallel to the machined surface (Figure 5.4(c)). Similar observations were also reported in an experimental study by Yanming *et al.* [YAN 03], who machined SiC particulate-reinforced MMC.

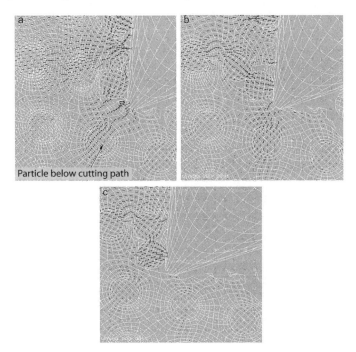

Figure 5.4. *Evolution of stress fields for a particle below the cutting path. Compressive and tensile stresses are represented by black >–< and white <—> symbols, respectively. Their lengths represent comparative magnitudes [PRA 07]*

5.4.2. *Development of the plastic zone*

The matrix in between particle and tool, and that at the upper part of the particle are highly strained (Figure 5.5(a)) when a particle is at the lower part of the cutting edge. With the progression of cutting, the tool interacts with the particle at the cutting edge and the particle is debonded. It then slides and indents (Figure 5.5(b) & (c)) into the new workpiece surface causing high plastic strain in the surrounding matrix. As the tool moves further, the particle is released from the matrix leaving a ploughed hole in the surface with high residual strain (Figure 5.5(d)).

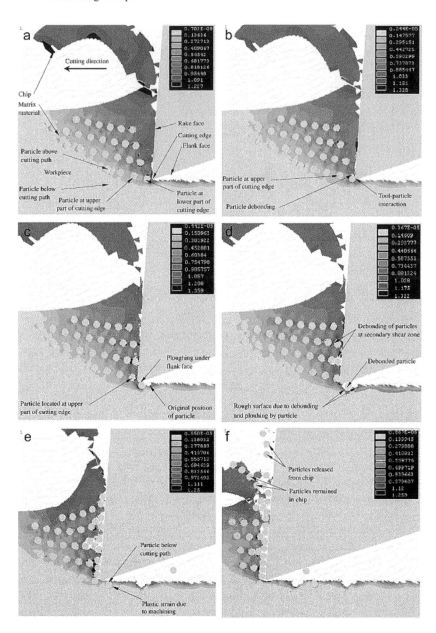

Figure 5.5. *Distribution of von Mises strain during machining of MMC [PRA 07]*

Particles located at the upper part of the cutting edge move up slightly with the advancement of the tool (Figure 5.5(c)). In this case, the strain in the matrix in between the particle and tool is not as high as the strain for a particle at the lower part of the cutting edge discussed earlier. The interaction between the particle and the tool is observed in this case with further progression of the tool (Figure 5.5(d)). The particle then partially debonds and slides along the rake face with the chip (Figure 5.5(e)).

Particles above the cutting path move in the cutting direction with the surrounding matrix due to the movement of the tool. As the rake face of the tool approaches, particle interface becomes highly strained (Figure 5.5(b)). Due to the ability of the matrix to deform plastically and particle's inability to do this, the matrix material experiences very high plastic strain. With further advancement of the tool, particles debond partially, interact with the tool and particles nearby, and move with the chip along the rake face (Figure 5.5(c)-(e)).

At the secondary deformation zone (tool-chip interface), the matrix experiences severe deformation, hence interfaces of most particles in the chip are highly strained. Additionally, most of the particles debond completely while passing through the secondary deformation zone (Figure 5.5(f)).

The interfaces of particles in the workpiece far below the cutting edge do not experience any plastic deformation due to machining. However, those situated immediately below the cutting edge are subjected to plastic deformation when the tool passes over them (Figure 5.6(e)). The banded pattern of the strain field is fragmented in the interface of particles just below the tool cutting edge. With further advancement of the tool, most of the interface of the particle is plastically deformed (Figure 5.5(e)). Additionally, the matrix at the matrix–tool cutting edge interface is plastically strained. The particles immediately below the cutting edge seem to act like indenters due to their interaction with the tool. In these regions the matrix can be seen to be plastically deformed to a greater depth (Figure 5.5(f)).

5.4.3. *Comparison of experimental and FE simulation observations*

It is of interest to note that some phenomena such as (i) the flow of particles in the chip root region, (ii) the partial debonding of particles from the matrix near the secondary shear zone, (iii) the continuous sliding of particles over the rake face, (iv) particle–cutting edge interaction and (v) the pull out of particles from machined surface observed in FEA can be explained from experimental results obtained by several investigators.

Hung *et al.* [HUN 99] reported (i) cracks due to debonding of particles in front of the tool and (ii) aligned reinforced particles along the shear plane in the chip root region. El-Gallab *et al.* [ELG 98b] observed the flow lines of particles and debonded particles in the chips. Almost all researchers noted comparatively high tool wear during machining of MMCs with any tool.

For diamond tools it is reported that abrasive wear at the rake face is smoother than that at the flank face [ELG 98a, DIN 05, HEA 01]. The smoother rake face wear can be attributed to frequent interactions between the rake face and hard particles, and the continuous sliding of these particles along the rake face (Figure 5.5(e) & (f)).

Several researchers [ELG 98a, DIN 05, CHA 96] have reported grooves and chipping (due to repeated impact between the tool edge and particles) on the cutting edge and flank face after machining MMCs. The damage of the tool cutting edge/flank was attributed to abrasion [ELG 98a, LIN 01, DAV 02] and pull out of tool material grains from cutting edge and flank face of the tool [CHA 96]. It was also reported that flank wear increases with the increase of speed [KIL 05, CIF 04], because at higher speeds impact between particle and tool increases which causes chipping of the cutting edge [CIF 04].

Due to interaction with the cutting edge, particles on the lower part of the cutting edge are debonded and pulled out leaving cavities on the machined surface. Zhang *et al.* [ZHA 95] and Yan *et al.* [YAN 95] who studied MMCs using scratching tests observed pull out of reinforcement particles and cavities on the scratched surface. Similar observations were also reported in an experimental study by El-Gallab *et al.* [ELG 98b] and Jaspers *et al.* [JAS 02] who machined SiC particulate-reinforced MMC.

5.5. Summary

Most of the studies in the conventional machining of MMCs are related to turning. Thus, a reasonable understanding of chip formation, tool wear and tool–particle–matrix interaction during turning is essential. Although several methods are used in the machining of metal matrix composites, the effect of reinforcement particles, such as high tool wear, cavities and work hardening in the machined surface, on machining performance is still unavoidable. Diamond is the most suitable tool material from the hardness point of view. Two-body and three-body abrasions and adhesion are mainly responsible for deteriorated tool performance. The surface quality obtained by a conventional machining process is mainly controlled by reinforcement size, feed and sharpness of a cutting edge.

Research in non-conventional machining of MMCs is still primary. Thus the mechanisms of most of the processes are not totally clear. Some relationships between input and output parameters have been established for non-conventional machining processes. Tool wear in electro-discharge machining, excessive heating in laser cutting, inaccuracy in abrasive jet machining and slow processing of electro-chemical machining remain the major problems to solve. It seems that in-depth investigations into non-conventional machining may bring about promising results in the processing of MMCs as these processes have some mentionable advantages over conventional machining processes.

5.6. References

[ABR 92] ABRATE S., WALTON D., "Machining of composite materials: Part II: Non-traditional methods", *Composites Manufacturing No. 2*, pp 85-94, 1992.

[AND 00] ANDREWES C.J.E., FENG H.Y., LAU W.M., "Machining of an aluminum/SiC composite using diamond inserts", *Journal of Materials Processing Technology*, vol. 102, 2000, p. 25–29.

[BAS 06] BASAVARAJAPPA S., CHANDRAMOHAN G., RAO K.V.N.,. RADHAKRISHANAN R., KRISHNARAJ V., "Turning of particulate metal matrix composites – review and discussion", *Proc. IMechE Part B: J. Engineering Manufacture*, vol. 220, 2006, p. 1189-1204.

[CAP 96] CAPELLO E., POLINI W., SEMERARO Q., "Abrasive water jet cutting of MMC: Analysis of the quality of the generated surfaces", *Proceedings of the 1996 3rd Biennial Joint Conference on Engineering Systems Design and Analysis*, ESDA. Part 7 (of 9), 1-4 July 1996.

[CHA 96] CHAMBERS A.R., "The machinability of light alloy MMCs", *Composites: Part A*, vol. 27, p. 143–147, 1996.

[CHA 99] CHANG S.T., TUAN W.H., YOU H.C., LIN I.C., "Effect of surface grinding on the strength of NiAl and Al_2O_3/NiAl composites", *Materials Chemistry and Physics*, vol. 59, 1999, p. 220-224.

[CIF 04] CIFTCI I., TURKER M., SEKAR U., "Evaluation of tool wear when machining SiC-reinforced Al-2014 alloy matrix composites", *Materials and Design*, vol. 25, 2004, p. 251–255.

[DAH 89] DAHOTRE N.B., MCCAY T.D., MCCAY M.H., "Laser processing of a SiC/Al-alloy metal matrix composite", *Journal of Applied Physics*, vol. 65 no. 12, 1989, p. 5072-5077.

[DAV 00] DAVIM J.P., BAPTISTA A.M., "Relationship between cutting force and PCD cutting tool wear in machining silicon carbide reinforced aluminium", *Journal of Materials Processing Technology*, vol. 103, 2000, p. 417–423.

[DAV 01] DAVIM J.P., ANTÓNIO C.A.C, "Optimal drilling of particulate metal matrix composites based on experimental and numerical procedures", *International Journal of Machine Tools and Manufacture*, vol. 41, no. 1, 2001, p. 21–31.

[DAV 02] DAVIM J.P., "Diamond tool performance in machining metal–matrix composites", *Journal of Materials Processing Technology*, vol. 128, 2002, p. 100–105.

[DAV 03] DAVIM J.P., "Study of drilling metal –matrix composites based on the Taguchi techniques", *Journal of Materials Processing Technology*, vol. 132, 2003, p. 250-254.

[DAV 07] DAVIM J.P., SILVA J., BAPTISTA A.M., "Experimental cutting model of metal matrix composites (MMCs)", *Journal of Materials Processing Technology*, vol. 183, 2007, p. 358–362.

[DER 01] D'ERRICO G.E., CALZAVARINI R., "Turning of metal matrix composites", *Journal of Materials Processing Technology*, vol. 119, 2001, p. 257–260.

[DIN 05] DING X., LIEW W.Y.H., LIU X.D., "Evaluation of machining performance of MMC with PCBN and PCD tools", *Wear*, vol. 259, 2005, p. 1225–1234.

[ELG 98a] EL-GALLAB M., SKLAD M., "Machining of Al/SiC particulate metal matrix composites, Part I: tool performance", *Journal of Material Processing Technology*, vol. 83, 1998, p. 151–158.

[ELG 98b] EL-GALLAB M., SKLAD M., "Machining of Al/SiC particulate metal matrix composites, Part II: work surface integrity", *Journal of Material Processing Technology*, vol. 83, 1998, p. 277–285.

[ELG 00] EL-GALLAB M., SKLAD M., "Machining of Al/SiC particulate metal matrix composites part III: comprehensive tool wear models comprehensive tool wear models", *Journal of Materials Processing Technology*, vol. 101, 2000, p. 10–20.

[HAS 95] HASHISH M., "Waterjet machining of advanced composites", *Materials and Manufacturing Processes*, vol. 10, no. 6, 1995, p. 1129-1152.

[HEA 01] HEATH P.J., "Developments in applications of PCD tooling", *Journal of Materials Processing Technology*, vol. 116, no. 1, 2001, p. 31–38.

[HIH 00] HIHARA L.H., PANQUITES P., Method of electrochemical machining (ECM) of particulate metal-matrix composites (MMCs), Patent no. 6110351, 29 August 2000.

[HIH 03] HIHARA L.H., PANQUITES P., "The potential of electrochemical machining for silicon carbide/aluminum metal-matrix composites", *Abrasives and Grinding Magazine*, December-January, p. 12-17, 2003.

[HOC 97] HOCHENG H., LEI W.T., HSU H.S., "Preliminary study of material removal in electrical-discharge machining of SiC/Al", *Journal of Materials Processing Technology*, vol. 63, 1997, p. 813-818.

[HON 97] HONG L., VILAR R.M., YOUMING W., "Laser beam processing of a SiC particulate reinforced 6061 aluminium metal matrix composite", *Journal of Materials Science*, vol. 32, 1997, p. 5545-5550.

[HOO 99] HOOPER R.M., HENSHALL J.L., KLOPFER A., "The wear of polycrystalline diamond tools used in the cutting of metal matrix composites", *International Journal of Refractory Metals & Hard Materials*, vol. 17, 1999, p. 103–109.

[HUN 94] HUNG N.P., YANG L.J., LEONG K.W., "Electrical discharge machining of cast metal matrix composites", *Journal of Materials Processing Technology*, vol. 44, 1994, p. 229–236.

[HUN 97] HUNG N.P., ZHONG Z.W., ZHONG C.H., "Grinding of Metal Matrix Composites Reinforced with Silicon-Carbide Particles", *Materials and Manufacturing Processes*, vol. 12, no. 6, 1997, p. 1075-1091.

[HUN 99] HUNG N.P., LOH N.L., VENKATESH V.C., "Machining of metal matrix composites", *Machining of Ceramics and Composites*, edited by S. JAHANMIR, M. RAMULU and P. KOSHY, 1999, Marcel Dekker, Inc., New York, Basel.

[ILI 96] DI ILIO A., PAOLETTI A., TAGLIAFERRI V., VENIALI F., "An experimental study on grinding of silicon carbide reinforced aluminium alloys", *International Journal of Machine Tools and Manufacture*, vol. 36, no. 6, 1996, p. 673-685.

[JAS 02] JASPERS S.P.F.C., DAUTZENBERG J.H., "Material behaviour in metal cutting: strains, strain rates and temperatures in chip formation", *Journal of Materials Processing Technology*, vol. 121, 2002, p. 123–135.

[JOS 99] JOSHI S.S., RAMAKRISHNAN N., NAGARWALLA H.E., RAMAKRISHNAN P., "Wear of rotary carbide tools in machining of Al/SiCp composites", *Wear*, vol. 230, 1999, p. 124–132.

[KAG 89] KAGAWA Y., UTSUNOMIYA S., KOGO Y., "Laser cutting of CVD-SiC fibre/A6061 composite", *Journal of Materials Science Letters*, vol. 8, 1989, p. 681-683.

[KIL 05] KILIÇKAP E., ÇAKIR O., AKSOY M., İNAN A., "Study of tool wear and surface roughness in machining of homogenised SiC-p reinforced aluminium metal matrix composite", *Journal of Materials Processing Technology*, vol. 164–165, 2005, p. 862–867.

[KAN 05] KANNAN S., KISHAWY H.A., DEIAB I.M., SURAPPA M.K., "Modeling of tool flank wear progression during orthogonal machining of metal matrix composites", *Trans. North Am. Manuf. Res. Inst. SME NAMRC*, vol. 33, 2005, p. 605–612.

[KAR 06] KARAKAŞ M.S., ACIR A., ÜBEYLI M., ÖGEL B., "Effect of cutting speed on tool performance in milling of B_4Cp reinforced aluminum metal matrix composites", *Journal of Materials Processing Technology*, vol. 178, 2006, p. 241–246.

[KIS 04] KISHAWY H.A., KANNAN S., BALAZINSKI M., "An energy based analytical force model for orthogonal cutting of metal matrix composites", *Annals of the CIRP*, vol. 53, 2004, p. 91–94.

[KIS 05] KISHAWY H.A., KANNAN S., BALAZINSKI M., "Analytical modeling of tool wear progression during turning particulate reinforced metal matrix composites", *Annals of CIRP*, vol. 54, no. 1, 2005, p. 55–58.

[KWA 08] KWAK J.S., KIM Y.S., "Mechanical properties and grinding performance on aluminum-based metal matrix composites", *Journal of Materials Processing Technology*, vol. 201, 2008, p. 596–600.

[LEE 95] LEE R.S., CHEN G.A., HWANG B.H., "Thermal and grinding induced residual stresses in a silicon carbide particle-reinforced aluminium metal matrix composite", *Composites*, vol. 26, 1995, p. 425-429.

[LIN 95] LIN J.T., BHATTACHARYYA D., LANE C., "Machinability of a silicon carbide reinforced aluminium metal matrix composite", *Wear*, vol. 181-183, 1995, p. 883-888.

[LIN 01] LIN C.B., HUNG Y.W., LIU W.-C., KANG S.-W., "Machining and fluidity of 356Al/SiC(p) composites", *Journal of Material Processing Technology*, vol. 110, 2001, p. 152–159.

[LOO 92] LOONEY L.A., MONAGHAN J.M., O'REILLY P., TAPLIN D.M.R., "The turning of an Al/SiC metal composite", *Journal of Materials Processing Technology*, vol. 33, 1992, p. 453–468.

[MAN 03] MANNA A., BHATTACHARAYYA B., "A study on machinability of Al-SiC MMC", *Journal of Materials Processing Technology*, vol. 140, 2003, p. 711–716.

[MAN 03b] MANNA A., BHATTACHARAYYA B., "Study on different tooling systems during turning for effective machining of Al-SiC MMC", *J. Prod. Eng. Inst. Eng.*, vol. 83, 2003, p. 46–50.

[MOH 02] MOHAN B., RAJADURAI A., SATYANARAYANA K.G., "Effect of SiC and rotation of electrode on electric discharge machining of Al–SiC composite", *Journal of Materials Processing Technology*, vol. 124, no. 3, 2002, p. 297-304.

[MOH 04] MOHAN B., RAJADURAI A., SATYANARAYANA K.G., "Electric discharge machining of Al–SiC metal matrix composites using rotary tube electrode", *Journal of Materials Processing Technology*, vol. 153–154, 2004, p. 978–985.

[MON 92] MONAGHAN J., O'REILLY P., "The drilling of an Al/SiC metal matrix composite", *Journal of Materials Processing Technology*, vol. 33, 1992, p. 469–480.

[MOR 95] MORIN E., MASOUNAVE J., LAUFER E.E., "Effect of drill wear on cutting forces in the drilling of metal-matrix composites", *Wear*, vol. 184, 1995, p. 11-16.

[MUB 94] MUBARKI B., FOWLE R.F., HEATH P.J., MATHEW P., BANDYOPADHYAY S., "Quantitative aspects of drilling a metal matrix composite", *Journal of the Australian Ceramic Society*, vol. 30, 1994, p. 137–163.

[MUB 95] MUBARKI B., BANDYOPADHYAY S., FOWLE R.F., MATHEW P., HEATH P.J., "Machining studies of an Al2O3–Al metal matrix composite. I. Drill wear characteristic", *Journal of Materials Science*, vol. 30, no. 24, 1995, p. 6273–6280.

[MÜL 00] MÜLLER F., MONAGHAN J., "Non-conventional machining of particle reinforced metal matrix composite", *International Journal of Machine Tools & Manufacture*, vol. 40, 2000, p. 1351–1366.

[MUT 08] MUTHUKRISHNAN N., MURUGAN M., RAO P.K., "Machinability issues in turning of Al-SiC (10p) metal matrix composites", *International Journal of Advanced Manufacturing Technology*, vol. 39, no. 3-4, 2008, p. 211-218.

[PED 05] PEDERSEN W.E., RAMULU M., "Proposed tool wear model for machining particle reinforced metal matrix composites", *Trans. North Am. Manuf. Res. Inst. SME Presented at NAMRC*, vol. 33, 2005, p. 549–556.

[PRA 06] PRAMANIK A, ZHANG L.C., ARSECULARATNE J.A., "Prediction of cutting forces in machining of metal matrix composites", *International Journal of Machine Tools and Manufacture*, vol. 46, 2006, p. 1795–8031.

[PRA 07] PRAMANIK A., ZHANG L.C., ARSECULARATNE J.A., "An FEM investigation into the behaviour of metal matrix composites: tool–particle interaction during orthogonal cutting", *International Journal of Machine Tools and Manufacture*, vol. 47, 2007, p. 1497–506.

[PRA 08] PRAMANIK A., ZHANG L.C., ARSECULARATNE J.A., "Deformation mechanisms of MMCs under indentation", *Composites Science and Technology*, vol. 68, 2008, p. 1304-1312.

[PRA 08b] PRAMANIK A., ZHANG L.C., ARSECULARATNE J.A., "Machining of metal matrix composites: Effect of ceramic particles on residual stress, surface roughness and chip formation", *International Journal of Machine Tools & Manufacture*, vol. 48, 2008, p. 1613–1625.

[PRA 08c] PRAMANIK A., ZHANG L.C., ARSECULARATNE J.A., "Machining of particulate-reinforced metal matrix composites", in *Machining: Fundamentals and Recent Advances*, edited by J. Paulo DAVIM, Springer London, 2008-07.

[PRA 08d] PRAMANIK A., Understanding the deformation and material removal mechanisms of particulate-reinforced metal matrix composites subjected to machining, PhD thesis, University of Sydney, 2008.

[RAM 89] RAMULU M., TAYA, M.I., "EDM machinability of SiCw/Aw composites", *Journal of Materials Science*, vol. 24, no. 3, 1989, p. 1103-1108.

[RAM 02] RAMULU M., RAO P.N., KAO H., "Drilling of $(Al_2O_3)_p$/6061 metal matrix composites", *Journal of Materials Processing Technology*, vol. 124, 2002, p. 244–254.

[RED 08] REDDY N.S.K., KWANG-SUP S., YANG M., "Experimental study of surface integrity during end milling of Al/SiC particulate metal–matrix composites", *Journal of Materials Processing Technology*, vol. 201, 2008, p. 574–579.

[RON 09] RONALD B.A., VIJAYARAGHAVAN L., KRISHNAMURTHY R., "Studies on the influence of grinding wheel bond material on the grindability of metal matrix composites", *Materials and Design*, vol. 30, 2009, p. 679–686.

[ROZ 01] ROZENEK M., KOZAK J., DĄBROWSKI L., ŁUBKOWSKI K., "Electrical discharge machining characteristics of metal matrix composites", *Journal of Materials Processing Technology*, vol. 109, 2001, p. 367-370.

[SAH 02] SAHIN Y., KOK M., CELIK H., "Tool wear and surface roughness of Al_2O_3 particle-reinforced aluminium alloy composites", *Journal of Materials Processing Technology*, vol. 124, 2002, p. 280–291.

[SEN 09] SENTHILKUMAR C., GANESAN G., KARTHIKEYAN R., "Study of electrochemical machining characteristics of Al/SiC_p composites", *International Journal of Advanced Manufacturing Technology*, DOI: 10.1007/s00170-008-1704-1, 2009.

[SHA 02] SHANMUGAM D.K., CHEN F.L., SIORES E., BRANDT M., "Comparative study of jetting machining technologies over laser machining technology for cutting composite materials", *Composite Structures*, vol. 57, 2002, p. 289–296.

[SIN 04] SINGH P.N., RAGHUKANDAN K., RATHINASABAPATHI M., PAI B.C., "Electric discharge machining of Al–10%SiC$_P$ as-cast metal matrix composites", *Journal of Materials Processing Technology*, vol. 155–156, 2004, p. 1653–1657.

[TOM 92] TOMAC N., TONNESSEN K., "Machinability of particulate aluminum matrix composites", *Annals of CIRP*, vol. 41, 1992, p. 55–58.

[TOS 04] TOSUN G., MURATOGLU M., "The drilling of Al/SiCp metal–matrix composites. Part II: workpiece surface integrity", *Composites Science and Technology*, vol. 64, 2004, p. 1413–1418.

[TOS 04a] TOSUN G., MURATOGLU M., "The drilling of Al/SiCp metal–matrix composites. Part I: microstructure", *Composites Science and Technology*, vol. 64, 2004, p. 1413–1418.

[ÜBE 07] ÜBEYLI M., ACIR A., KARAKAŞ M.S., ÖGEL B., "Study on performance of uncoated and coated tools in milling of Al-4%Cu/B4C metal matrix composites", *Materials Science and Technology*, vol. 23, no. 8, 2007, p. 945-950.

[ÜBE 08] ÜBEYLI M., ACIR A., KARAKAŞ M.S., ÖGEL B., "Effect of feed rate on tool wear in milling of Al-4%Cu/B$_4$C$_p$ composite", *Materials and Manufacturing Processes*, vol. 23, 2008, p. 865–870.

[WAN 00] WANG C.C., YAN B.H., "Blind-hole drilling of Al$_2$O$_3$/6061Al composite using rotary electro-discharge machining", *Journal of Materials Processing Technology*, vol. 102, 2000, p. 90-102.

[WAN 03] WANG J., HUNG C.Z., SONG W.G., "The effect of tool flank wear on the orthogonal cutting process and its practical implications", *Journal of Materials Processing Technology*, vol. 142, 2003, p. 338–346.

[WEI 93] WEINERT K., "A consideration of tool wear mechanism when machining metal matrix composite (MMC)", *Annals of CIRP*, vol. 42, 1993, p. 95–98.

[XIA 01] XIAOPING L., SEAH W.K.H., "Tool wear acceleration in relation to workpiece reinforcement percentage in cutting of metal matrix composites", *Wear*, vol. 247, 2001, p. 161–171.

[YAN 95] YAN C., ZHANG L.C., "Single-point scratching of 6061 Al alloy reinforced by different ceramic particles", *Applied Composite Materials*, vol.1, 1995, p. 431–447.

[YAN 99] YAN B.H., WANG C.C., "The machining characteristics of Al_2O_3/6061Al composite using rotary electro-discharge machining with a tube electrode", *Journal of Materials Processing Technology*, vol. 95, 1999, p. 222-231.

[YAN 00] YANMING Q., ZEHUA Z., "Tool wear and its mechanism for cutting SiC particle reinforced aluminium matrix composites", *Journal of Materials Processing Technology*, vol. 100, no. 1, 2000, p. 194–199.

[YAN 03] YANMING Q., BANGYAN Y., "The effect of machining on the surface properties of SiC/Al composites", *Journal of Materials Processing Technology*, vol. 138, no. 1–3, 2003, p. 464–467.

[ZHA 95a] ZHANG Z.F., ZHANG L.C., MAI Y.W., "Wear of ceramic particle-reinforced metal-matrix composites, part I wear mechanisms", *Journal of Materials Science*, vol. 30, no. 8, 1995, p. 1961–1966.

[ZHA 95b] ZHANG Z.F., ZHANG L.C., MAI Y.W., "Wear of ceramic particle-reinforced metal–matrix composites, part II a model of adhesive wear", *Journal of Materials Science*, vol. 30, no. 8, 1995, p. 1967–1971.

[ZHA 95c] ZHANG Z.F., ZHANG L.C., MAI Y.W., "Particle effects on friction and wear of aluminium matrix composites", *Journal of Materials Science*, vol. 30, no. 23, 1995, p. 5999–6004.

[ZHA 97] ZHANG Z.F., ZHANG L.C., MAI Y.W., "Modeling steady wear of steel/Al2O3– Al particle reinforced composite system", *Wear*, vol. 211, no. 2, 1997, p. 147–150.

[ZHO 00] ZHONG Z., HUNG N.P., "Diamond turning and grinding of aluminum-based metal matrix composites", *Materials and Manufacturing Processes*, vol. 15, no. 6, 2000, p. 853-865.

[ZHO 02] ZHONG Z., HUNG N.P., "Grinding of alumina/aluminum composites", *Journal of Materials Processing Technology*, vol. 123, no. 1, 2002, p. 13-17.

[ZHO 03] ZHONG Z.W., "Grinding of aluminium-based metal matrix composites reinforced with Al_2O_3 or SiC Particles", *International Journal of Advanced Manufacturing Technology*, vol. 21, 2003, p. 79–83.

Chapter 6

Machining Ceramic Matrix Composites

6.1. Introduction

Ceramic matrix composites (CMC) are a very important class of materials especially in the area of machining technology as they are used extensively for manufacturing cutting tool inserts. The machining of these materials presents manufacturing engineers with a very difficult task owing to their high hardness and extreme mechanical and physical properties. This chapter provides a current review of the state of the art of machining ceramic matrix composite materials and includes electro-discharge machining, laser machining, ultrasonic machining, grinding, and also provides a narrative on the application of these techniques to manufacturing cutting tool inserts.

6.2. Electro-discharge machining of CMCs

According to Put *et al.* [PUT 01] the popularity of zirconia-based ceramic composites has been growing in recent years, leading to a 6-8% share of the technical ceramics market. Components made from

---

Chapter written by Mark J. JACKSON and Tamara NOVAKOV.

zirconia-based ceramic composites can be found in pump components, paper cutting tools, automotive parts, heat insulators, exhaust liners, and biomaterials. Put *et al.* [PUT 01] deals with the shaping of zirconia-based composites with 30 vol.% and 40 vol.% TiB_2, TiN, TiC and TiCN, as secondary phases using die sink EDM, currently considered to be the most advanced precision ceramic machining technology. The process was evaluated using volumetric material removal rate, surface finish and tool wear while the process parameters were kept constant. Changing the polarity influenced optimum parameters of the process. The influence of optimized and non-optimized conditions has been discussed.

It has been stated that the polarity had an influence on different composites. When using positive polarity, the volumetric material removal rate (VMRR) was higher for ZrO_2–TiN, ZrO_2–TiCN and ZrO_2–TiC composites but it was lower for the ZrO_2–TiB_2 composites. Although the VMRR was higher for positive polarity it has also been shown that it induces less stable machining conditions with the possibility of sporadic arcing. Apart from ZrO_2–TiC and ZrO_2–TiCN under negative polarity, the higher percentage of conductive phases yielded higher VMRR. The influence of polarity on VMRR for CMCs is shown in Figure 6.1 [PUT 01].

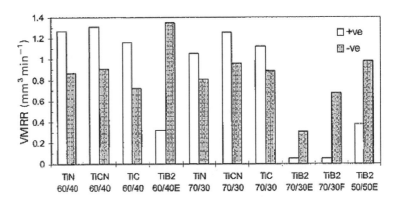

Figure 6.1. *VMRR for ZrO_2 based composites machined using EDM with positive and negative polarities [PUT 01]*

When the tool is under negative polarity the tool wear is shown to be higher probably due to the fact that the positive ions which impart onto the tool are heavier than the negative ions which impart when the polarity is positive.

Surface roughness was shown to be dependent on the polarity used, due to the fact that polarity influences the VMRR and process stability. Since the VMRR was lower using negative polarity and the process had less sporadic arcing, the surface roughness in this case was shown to be better than when positive polarity was used (Figure 6.2) [PUT 01].

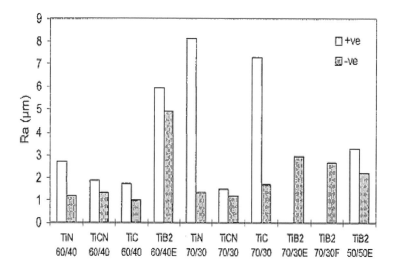

Figure 6.2. *Surface roughness for ZrO_2 based composites machined using EDM with positive and negative polarities [PUT 01]*

When the effect of current values was investigated it was concluded that due to the fact that VMRR increases with the increase in current, the surface roughness deteriorated (Figure 6.3).

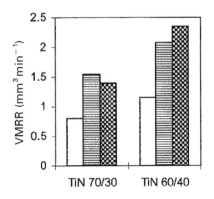
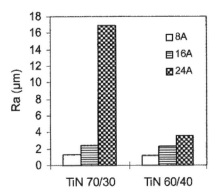

Figure 6.3. *Influence of current on VMRR and surface roughness of ZrO_2-TiN composites [PUT 01]*

The fine surface finish was possible only when using low current. The duty factor and the pulse interval was shown to have a significant influence on the process, showing that the increase in the duty factor yields a deterioration in the surface roughness (Figure 6.4) [PUT 01].

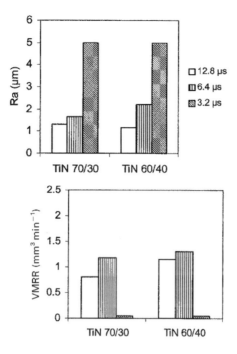

Figure 6.4. *Influence of duty factor on VMRR and surface roughness of ZrO_2-TiN composites [PUT 01]*

When material properties were investigated, it was concluded that there is a 25% drop in bending strength compared to sawn specimens, but the bending strength was 800 MPa. Silicon nitride-based composites have caused interest for a long time due to their superior mechanical characteristics such as high hardness and high wear resistance. Characteristics like these have made them attractive in the bearing and cutting tool industries. Even though their characteristics are superior, their cost was high due to the fact that diamond tools had to be used in order to obtain quality parts. Even then, the process is difficult and the toughness of the final product was alarmingly low. EDM is a technology that can be applied to material of any hardness as long as it is conductive with a good surface finish and has no contact with the tool [JON 01]. 40% TiB_2 has been added to the matrix in order to provide conductivity of the material and therefore enable the use of EDM. The research sought to identify the optimal machining parameters that yield stable machining conditions. Both die sink and wire EDM have been discussed, showing machinability using the volumetric removal rate (VMRR), surface finish and relative volumetric tool wear (RVTW), effects of polarity, current and pulse length. Three different combinations of parameters have been used to represent extreme, rough, and fine machining regimes. Damage to the surface has been investigated using a scanning electron microscope. VMRR has been shown to be higher with negative polarity (Figure 6.5a).

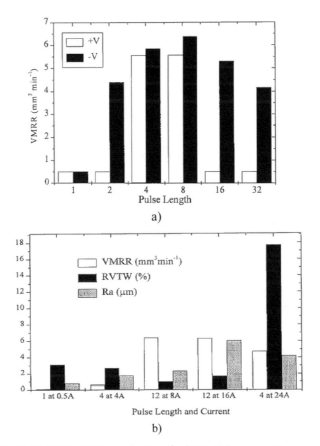

Figure 6.5. *(a) VMRR for EDM machining of a Si_3N_4- TiB_2 composite using different polarities and lengths; (b) VMRR, RVTW and Ra for a Si_3N_4- TiB_2 composite using relative pulse lengths and currents under negative polarity [JON 01]*

Negative polarity EDM was also shown to be more stable and sustainable over a wider range of pulse durations. When investigating die sink EDM using negative polarity, the pulse duration was shown to have a dramatic impact on RVTW (Figure 6.5b). RVTW increased with lower pulse duration leaving surface roughness at constant values. When using WEDM (Figure 6.6), VMRR increased with increasing pulse length while the surface roughness was kept constant regardless of the applied pulse duration. No cracks were reported in the re-cast layer at any current values, indicating no destruction of the surface layer.

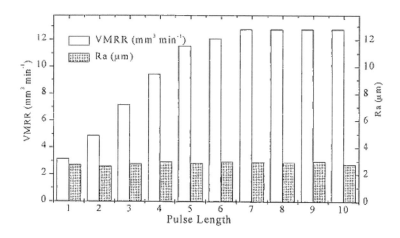

Figure 6.6. *VMRR and Ra for WEDM machining of Si_3N_4- TiB_2 using 350 A current [JON 01]*

Fu *et al.* [FU 94] considers the influence of EDM on the strength of an Al_2O_3 with 40 vol.% Cr_3C_2 composite. Operating parameters have been investigated and it has been concluded that the pulse length has a great influence on strength. Depending on the pulse current, two separate material removal processes have been determined: melting at low pulses and thermal spalling (removal of larger particles), and melting at high pulses [FU 94].

Surface characterization has been performed using SEM, Auger electron spectroscopy and EDAX (x-ray energy dispersive analysis).

Surface roughness has been analyzed using a profilometer while bending strength was used as the value for determining surface damage. With the increase in pulse current to over 0.8 A, a drop in strength occurred followed by an increase in surface roughness (Figure 6.7) [FU 94].

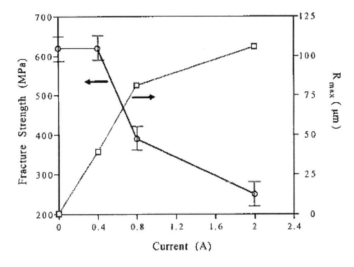

Figure 6.7. *Fracture strength and surface roughness of EDM specimens for various pulse currents [FU 94]*

Surface morphology had undergone changes when comparing pulse currents of 0.4 and 2 A (Figure 6.8). At 0.4 A, Cr_3C_2 has melted away, leaving Al_2O_3 intact, while at higher pulses, vaporization has been detected leaving craters, spheroids and cracks in the material. By increasing pulse current, groove-like defects act as stress concentrators initiating crack formation and propagation leading to failure, resulting in a sudden drop in strength values. Owing to ceramic brittleness and hardness they are prone to failure due to temperature changes. This is inevitable with EDM especially with higher pulse currents and thus leads to rough surface finish spalling beneath the melting zone. At low pulse currents spalling is avoided with melting taking its place, resulting in a smaller thermal gradient and good surface finish. After looking at the surface of the material subjected to the plasma stream during EDM, it was determined that there is crack formation in the

EDM phase created by melting and solidification and shrinkage. The re-solidified layer consisted of the same material as the composite without the creation of solid solutions or chemical reactions between phases. Owing to the melting of the phases, a dendritic structure forms in the Al_2O_3 phase possibly due to molten material re-solidifying in the direction of the thermal gradient.

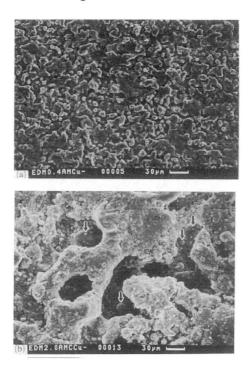

Figure 6.8. *Surface morphology of an EDM face with pulse currents of a) 0.4 A and b) 2.0 A [FU 94]*

Pitman and Huddleston [PIT 00] deal with sinking electrical discharge machining (EDM) of a zirconia-based ceramic matrix composite reinforced with TiN dispersoid. TiN enables conduction and is necessary for EDM application. VMRR and RVTW have been analyzed under induced arcing and normal sparking conditions. Material removal rate (MRR) mechanisms have been determined and discussed. It can be seen from Figure 6.9 that there are maximum values of VMRR that can be achieved at a certain duty factor after which there was a sharp decrease in the VMRR values. The RVTW values have dropped significantly when moving away from the lower duty factor value. However, it can also be seen that they are negative values. This has been explained by the build-up of carbon on the tool due to degradation of the dielectric.

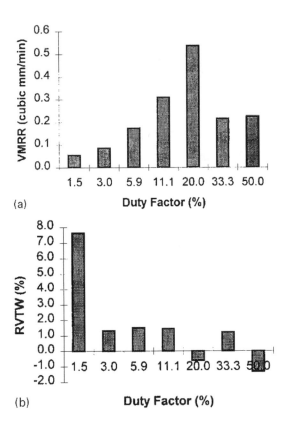

Figure 6.9. *Effect on VMRR and RVTW of increasing duty factor with medium discharge duration of 50 ms [PIT 00]*

When the influence of discharge duration was analyzed (Figure 6.10) it was noted that there is an initial decrease in VMRR at the lower values of discharge duration, but the values were constant afterwards. It was stated that this could be a consequence of the time needed to reach temperature equilibrium. The influence of the duration on RVTW has a decreasing trend with an occasional anomalous result.

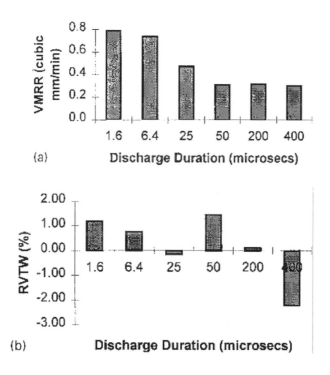

Figure 6.10. *Influence of discharge duration on VMRR and RVTW [PIT 00]*

The specific volumetric power consumption (SVPC) has been analyzed to reveal the cost efficient machining set of parameters. When looking at the influence of duty factor on SVPC for a constant value of the spark duration, no apparent change was detected. However, it has been determined that the values of SVPC are higher for higher values of discharge duration. Therefore, shorter discharge durations yield higher machining efficiencies. Increases in machining current at high discharge durations yielded a significant decrease in SPVC values, while this was not apparent for lower durations. When full arcing is applied, the SVPC values decrease. With the increase in duty factor, the values of SVPC for full arcing were higher than those for good arcing. This resulted in the fact that cost efficient material removal for arcing could be achieved by using low duty factors, short durations and higher current values [PIT 00]. Material removal mechanisms have been analyzed and it has been concluded that spalling and the removal of material in large segmented bowl-shaped particles was present (Figure 6.11).

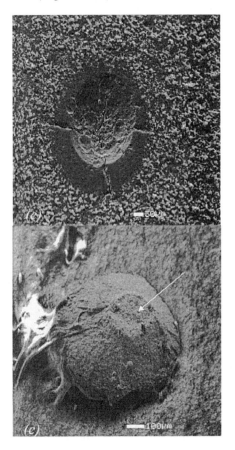

Figure 6.11. *Fracture pattern after light polish and fracture debris [PIT 00]*

Penny crack formation can be considered as a consequence of the rapid increase in temperature, while "quench cracks" and circumferential cracks are considered to be the result of the rapid cooling process. Lauwers *et al.* [LAU 08] considers the influence of the type and size of second phase particles on the EDM process parameters and surface roughness. ZrO_2 composite ceramics with WC, TiC and TiCN in a 60/40 vol.% composition of micro and nano particle sizes have been investigated. Test results for ZrO_2 –WC showed that cutting speed can be controlled through variation of discharge duration (te) and pulse interval time maintaining constant surface roughness. Both cutting speed and surface roughness are influenced by grain size with smaller grain size yielding lower cutting speeds and better surface roughness. Test results for ZrO_2 –TiCN showed opposite effects of the grain size on cutting speed showing an increase of cutting speed with a decrease in grain size as well as negligible effects on surface roughness. Test results for ZrO_2 –TiC show increases in surface roughness with decreases in grain size [LAU 08].

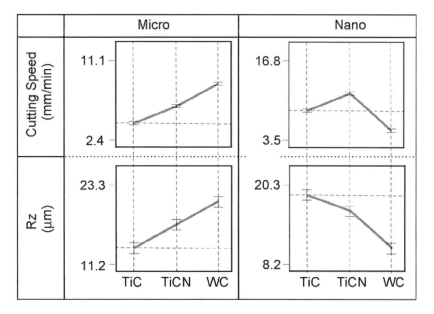

Figure 6.12. *Effect of the conductive phase [LAU 08]*

The size of the particles was shown to have a different effect on cutting speed for the different second phases added (Figure 6.12). It has been determined that there is an increase in cutting speed with the decrease in particle size from micro to nano for TiC and TiCN and vice versa for WC. The effects of the cutting speed and surface roughness are considered to be the result of different material removal mechanisms starting from melting to spalling and various chemical reactions. As mentioned previously, spalling occurs due to micro-cracks in the material generated due to the effect of the thermal gradient on materials with low toughness and low strength.

It has been concluded that there is an increase in cutting speed due to the lower thermal conductivity for the materials with melting as the primary material removal mechanism. If the material removal mechanism is driven by chemical reactions, the behavior is different and shows an increase in cutting speed with an increase in WC grain size. Defining primary material removal mechanisms enables the user to determine optimal parameters of EDM for various materials of interest [LAU 08].

6.3. Water jet machining of CMCs

Gonczy *et al.* [GON 94] provided an analysis that deals with the influence of water jet machining (WJM) on continuous fiber-reinforced ceramic composites. Effects on the surface roughness, tensile strength and strain failure have been investigated and compared to diamond machining of tensile test specimens.

Abrasive WJM is thought to have the best combination of parameters such as speed, heating, availability and cost compared to other technologies applied for machining of CMCs. Differences in tensile properties, dimensional tolerances, and surface finish have been discussed.

Specimens were cut using diamond wheels and WJM and were in ceramic and polymer materials. Analysis of dimensional tolerances showed that there were differences between diamond cut and WJM specimens. The diamond cut specimens had a lower coefficient of variation compared to WJM specimens yielding better tolerances. Also, one of the sets of specimens machined by WJM has been measured to be out-of-tolerance. Surface finish measurement showed distinctive differences between the four sets of tensile bars. Diamond cut ceramic had the best finish, followed by WJM ceramic, diamond cut polymer and WJM polymer. Regardless of the specimen types, diamond machining with the final grinding possesses a better surface condition than WJM surfaces.

From scanning electron microscopy (SEM) images it can be seen that for the samples cut in the polymer condition there is a slight difference in the morphology showing a slightly smoother surface with diamond cutting compared to WJM.

For samples cut in the ceramic condition, diamond cutting produces considerably smoother surfaces compared to WJM. There were no great differences in ultimate tensile strength or failure strain for the samples. There was no negative effect on the tensile properties when WJM had been used. Differences in mechanical strength have been found to be statistically insignificant [GON 94].

6.4. Laser machining of CMCs

The paper by Tuersley *et al.* [TUE 98] is the first part of two papers that deal with the optimal processing parameters of Nd YAG lasers as the laser pulse parameters such as pulse energy, duration, and intensity in order to achieve the optimum material characteristics such as material removal rate and surface quality of CMCs. Comparisons with other types of ceramic matrix composites like borosilicate glass matrix composite and magnesium-alumino-silicate (MAS) glass-ceramic matrix composite (GCMC) have been made, where both materials have the same incorporated SiC fibers.

It has been stated that the investigated material showed enhancement in both cut rate and surface quality. The material investigated in the study is SiC fiber-reinforced SiC which has been CVD/CVI (chemical vapor deposition/chemical vapor infiltration) deposited.

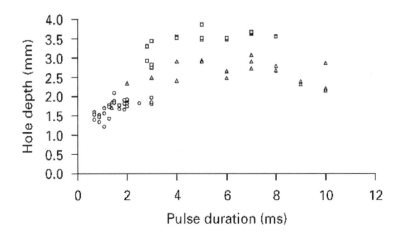

Figure 6.13. *Effect of pulse duration at constant pulse energies of:*
( ) 14.5 J, (Δ) 7.2 J and (O) 3.15 J [TUE 98]

Figure 6.13 shows an increase in material removal efficiency with the increase in pulse energy with the overlapping of all data sets at 3 ms. There is a general increase in penetration with a penetration limit at 7.2 J. It has been stated that pulse energy has a stronger influence on Material Removal Rate (MRR) than pulse width, which is in agreement with previous data.

Penetration capability of the investigated material is shown to be higher than that of glass matrix composites (GMC) and GCMC materials by 50-75%, although probably due to the higher amount of random porosity of the material, the scatter in the results is considered to be higher.

An investigation of the pulse energy variation has been presented with the results shown in Figure 6.14. It can be seen that there is an over 100% increase in material removal efficiency when compared to other materials.

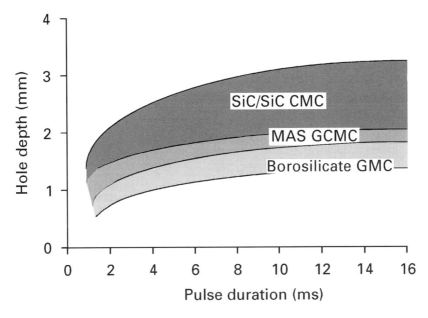

Figure 6.14. *A comparison of the effects of the pulse energy at constant pulse durations for the investigated materials [TUE 98]*

An investigation of the effect of varying peak power has been conducted with the results showing that the there is an increase in penetration with the increase in pulse energy. However, the change in the pulse duration only yields great effect if it is done from the lowest levels (Figure 6.15).

An investigation of the presence of different hole geometry or material damage due to varying the pulse duration, energy and intensity has been performed. It has been stated that there is a difference in the shape of drilled holes for the three materials. For the GMC, the holes were tapered, but upon second penetration, resulted in parallel bore. For MAS matrix GCMCs the results were similar but the bottom was more pointed. The SiC/SiC material had a convergent taper. In the SiC/SiC the tapered shape can be due to removal of the matrix instead of secondary phases, and the vaporization front could have the Gaussian profile of the energy distribution instead of the conduction-induced circular shape.

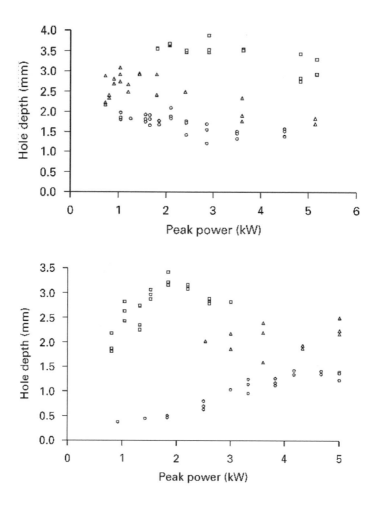

Figure 6.15. *Peak power versus hole depth for varied pulse durations and energies: ( ) 14.5 J, (Δ) 7.2 J and (O) 3.15 J [TUE 98]*

The amount of re-deposited material is less in the case of CMCs on the surface around the entry or on the bore of the hole (Figure 6.16).

The countersink feature that was present in the rest of the materials is absent in SiC/SiC [TUE 98].

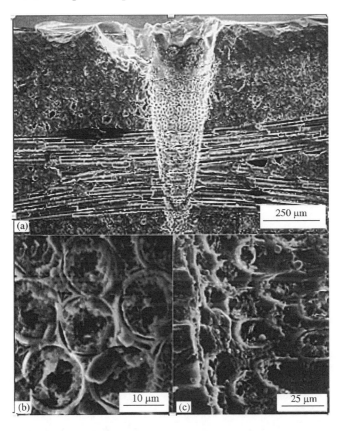

Figure 6.16. *(a) General geometry of hole: note the lack of re-deposited material in the bore of the hole and the cleanliness of the entrance hole; (b) through-ply and (c) cross-ply fibers exposed by the processed hole. Note the evidence of the pure carbon interfacial layer around each fiber [TUE 98]*

Tuersley *et al.* [TUE 98] focuses on parameters such as the choice and pressure of assisting gas and the point of focus of the incident laser beam and how they affect surface quality. As in the previous study, the comparison has been made with other composite materials. Variation of the focal point of the laser has been investigated with the aim of determining the effect on the materials in question (Figure

6.17). It can be seen that the results can explain the combinations of entry and exit hole diameters, explaining the shape of the obtained holes.

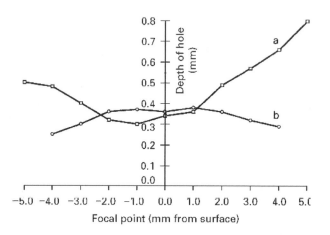

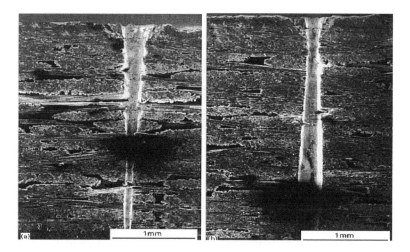

Figure 6.17. *Focal point trials (plate 3 mm thick, 5 bar assisting gas). Curve a entry hole diameter; curve b exit hole diameter. Photographs of (a) with the beam focused at a point 5 mm above the material surface; (b) with the beam focused at the material surface [TUE 98]*

When investigating maximum cutting speed, it has been shown that the cut speed is inversely proportional to material thickness (Figure 6.18) with a 30% increase in maximum cutting speed compared to MAS matrix material.

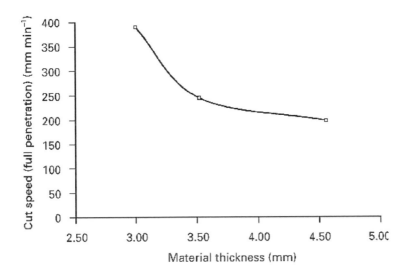

Figure 6.18. *Maximum cut speed (3.5 kW) versus material thickness (5 bar N2 assisting gas) [TUE 98]*

When using different assisting gases, it has been determined that there is almost no silicate re-deposited with nitrogen, argon or air assisting gas, and very little when oxygen is used. Therefore, the oxidation potential of SiC is considered to be low. However, the possibility of exposure of the CVD interfacial carbon layer does become an issue. This would mean that the adhesion between the matrix and fibers would be highly significant for the mechanical characteristics of the material.

Laser machining of SiC/SiC woven composites and SiN/BN fibrous monolithic composites has been investigated. The heat-affected zone has been shown to be 100-200 μm in depth and limited mostly to the matrix material. Different combinations of speed and focal point of the laser yielded different results. When using 5 and 10 cm/sec with the laser pointed at the root of the cut, more condensed SiC was observed due to the higher amount of energy absorbed in the cut area. Higher speed showed better surface roughness compared to lower speeds. The average surface roughness is considered to be close to the as-received material.

When laser machining SiN/BN it has been determined that the surface roughness does increase if multiple layers are removed at the same time, but when single layers are machined one at a time, surface roughness is lower than the samples produced by fused deposition modeling (FDM) [CAR 00].

6.5. Ultrasonic machining of CMCs

Hocheng *et al.* [HOC 00] explains ultrasonic drilling of 2D carbon-reinforced silicon carbide (C/SiC). Machining parameters of the drilling process such as abrasives, volume ratio, electric current and down-force have been investigated and their influence on MRR, hole clearance edge quality and tool wear has been presented. Ceramic matrix composites have been obtained through pyrolysis of polysilazane, and have been reinforced by woven carbon cloth. For the density enhancement, heat treatment has been applied. MRR has been shown to be proportionate to the static load applied on the tool. Owing to the higher impact force applied to the abrasive grains, more material can be removed from the surface.

Figure 6.19 shows the effect of grain size on material removal rate as well as hole clearance. It can be stated that there is an optimal size of the grains that yield a maximum value of the MRR. This can also be dependent on the amplitude of the tool oscillation. It has also been shown that the increase in grain size yields an increase in diameter clearance between the hole and the cutting tool.

The influence of various grains on MRR and hole clearance is presented in Figure 6.20, showing that the influence is dictated by the hardness of the grains and therefore the hardest B_4C grains have the highest influence on the MRR, while there is no significant influence of the grain type on hole clearance. The hardest grains also yield the best edge quality due to the fact that they cut the material in the neatest manner.

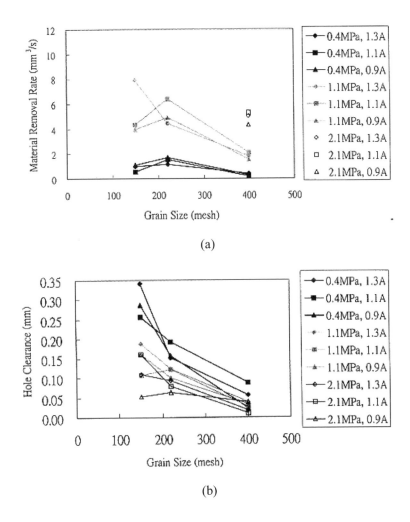

Figure 6.19. *Influence of various grain size on (a) material removal rate; (b) hole clearance [HOC 00]*

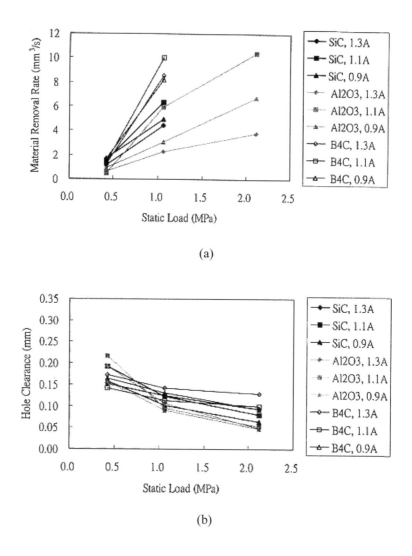

Figure 6.20. *Influence of various grains on a) material removal rate, b) hole clearance [HOC 00]*

When investigating tool wear, it has been determined that the tool wear is only 38.9 μm per hole and therefore can be considered to have good machinability compared to conventional drilling of carbide composites.

Owing to the results presented, the ultrasonic drilling process can be very competitive when compared to conventional drilling and non-conventional machining processes such as water jet machining which has issues with delamination and laser machining, which has problems with thermal stresses and heat-affected zones. When cost is analyzed, ultrasonic drilling can perform simultaneous multi-hole drilling and is considered to be more economical when compared to other processes.

Jianxin *et al.* [JIA 02] shows how the influence of workpiece material characteristics on MRR and surface integrity. Materials used for the study are: Al_2O_3/TiC, Al_2O_3/TiB_2 and $Al_2O_3/(Ti,W)C$ which are of three particle-reinforced ceramic composites, $Al_2O_3/SiCw$ is a whisker-reinforced ceramic composite, while $Al_2O_3/TiB_2/SiCw$ is toughened by the incorporation of both particles and whiskers.

The mechanical properties of the materials have been presented in Table 6.1. Ultrasonic machining has been conducted [JIA 02].

Code name	Compositions (vol.%)	Flexural strength (MPa)	Fracture toughness (MPa.m$^{1/2}$)
A	Al_2O_3/45% (Ti, W)C	800	4.9
B	Al_2O_3/55% TiC	900	5.04
C	Al_2O_3/25% TiB_2	780	5.2
D	Al_2O_3/20% TiB_2/ 10% SiC_w	750	7.8
E	Al_2O_3/30% SiC_w	760	8.6

Table 6.1. *Composition and mechanical properties of CMCs [JIA 02]*

When MRR was investigated, the effect of the fracture toughness showed that the composites with high fracture toughness had lower values of MRR, with whisker-reinforced ceramic composites having smaller MRR, and particle-reinforced ceramic composites having higher MRR (Figure 6.21).

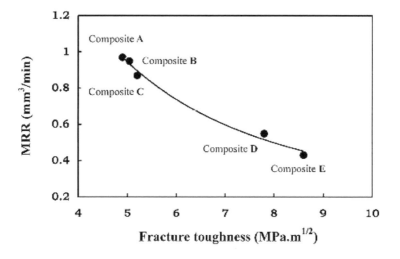

Figure 6.21. *Material removal rate in ultrasonic machining of different alumina-based ceramic composites [JIA 02]*

Surface roughness has been shown to be the highest for materials that yield the highest MRR (Figure 6.22). When microstructures of the material is observed it can be noted that mechanical characteristics depend on the whisker distribution in whisker-reinforced ceramics where the orientation of the whiskers yields a great influence.

Figure 6.23 shows the influence of the direction angle of the whiskers on MRR and surface roughness.

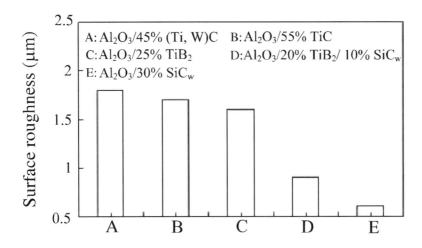

Figure 6.22. *Machined surface roughness in ultrasonic machining of different alumina-based ceramic composites [JIA 02]*

It can be stated that the effect of the whisker direction angle has similar trends for MRR and surface roughness (Figure 6.22). It can be seen that there is a decrease in MRR and surface roughness with the increase of the direction angle of whiskers. Influence of the whisker angle has also been seen on the cracks induced by particle impact during the ultrasonic machining process.

For lower angle values it has been seen that the whiskers are parallel to the machining surface and the lateral crack could propagate with ease through the matrix. Owing to this fact, the MRR of these materials is higher.

For 90° angles, whiskers were aligned normal to the machining surface and the crack propagation was resisted with whiskers acting as toughening mechanisms. Due to toughening, the MRR was low for these materials.

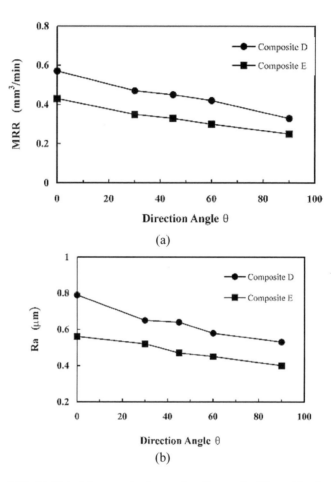

Figure 6.23. *(a) Material removal rates vs. direction angle, (b) machined surface roughness vs. direction angle in ultrasonic machining of whisker-reinforced ceramic composites [JIA 02]*

When investigating the strength of materials, the two-parameter Weibull distribution was performed (Figure 6.24). As stated in [JIA 02], in the Weibull distribution, a high value of Weibull modulus shows that the material in question has a high degree of homogenity of properties while a low Weibull modulus indicates that the surface damage during the final machining could have had some sort of influence.

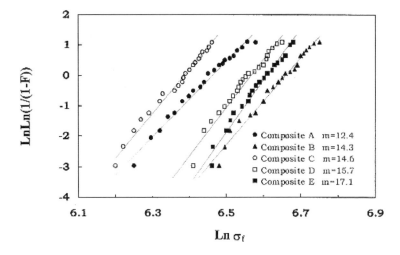

Figure 6.24. *Weibull plot of the strength distribution of the ultrasonic machined alumina-based ceramic composites [JIA 02]*

When compared, the whisker-reinforced ceramics yielded a higher Weibull module than particle-reinforced ceramics, implicating that composites with high fracture toughness have a higher Weibull modulus and therefore have a high degree of homogenity of properties [JIA 02]. The feasibility of using rotating ultrasonic machining (RUM) for machining of CMC has been investigated by Li *et al.* [LI 05]. MRR and cutting forces have been analyzed and chipping has also been observed. The results have been compared to those of drilled samples without the ultrasonic effect. In order to determine the effect of parameters such as spindle speed, feedrate and ultrasonic power, a three variable, two-level, full factorial design has been conducted with the output in the form of cutting force, MRR, and chipping dimensions.

A drill with metal-bonded diamond abrasives is vibrated using an ultrasonic mechanism in the axial direction and it is fed at a constant feedrate/force towards the workpiece material. The coolant is used to clear the hole, prevent jamming and cooling of the surface. It can be seen from Figure 6.25 that there is a 10% increase in the MRR for RUM compared to conventional diamond drilling operations. This is believed to be the consequence of the fact that the vibration induced by ultrasonic drilling increases the dynamic forces and flushes away the debris more effectively. Cutting forces have been averaged and presented in Figure 6.25. The biggest issue with cutting forces is the fact that due to the lamellar structure of the CMC and the presence of inclusions, the cutting forces deviate non-linearly. The cutting forces are dependent on feed rate and an increase in feed rate yields increase in cutting forces (Figure 6.26).

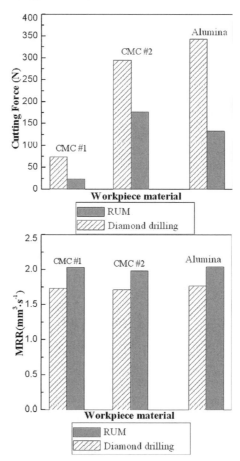

Figure 6.25. *Comparison of cutting forces and MRR in RUM and diamond drilling [JIA 02]*

When investigating surface quality by observing chipping of the material, it has been concluded that chipping in CMCs can be avoided by better adjustment of cutting parameters, or the use of sharper tools.

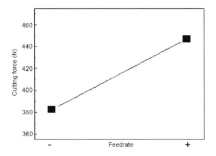

Figure 6.26. *Effect of feedrate on average cutting forces [JIA 02]*

When investigating MRR it has been shown that the increase of spindle speed, feed rate and ultrasonic power yield an increase in MRR (Figure 6.27).

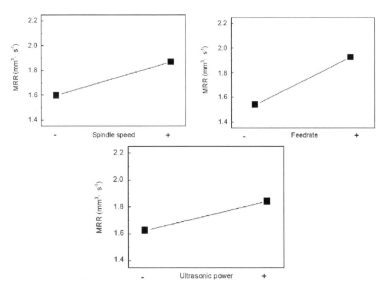

Figure 6.27. *Effects of spindle speed, feed rate and ultrasonic power on MRR [JIA 02]*

When hole quality is analyzed, due to the differences in material behavior, it is not enough to determine roundness, parallelism, and roughness but it is necessary to define chipping size and thickness. Chipping thickness and chipping size are proven to be inversely proportional to spindle speed, indicating that higher spindle speeds would yield a better surface quality in CMCs. The chipping thickness will be influenced by feedrate as well due to the fact that there is an interaction between MRR and feed.

The paper by Lee *et al.* [LEE 01] focuses on the effect of orientation and the amount of SiC whiskers in whisker-reinforced alumina. Material removal rate as well as surface roughness have been investigated at the 0 and 90 degree angles and evaluated for various amounts of reinforcing whiskers. Materials have been defined as : A – 10 vol.% of SiC, B – 20 vol.% of SiC, C – 30 vol.% of SiC. Figure 6.28 shows the influence of the direction angle on MRR and Ra for materials with three different amounts of whiskers.

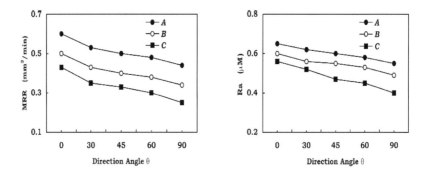

Figure 6.28. *Influence of the direction angle on MRR and Ra [LEE 01]*

It can be stated that the increase in the direction angle yields decreases in MRR. The effect is considered to be longer for the material with a higher amount of whisker content. The surface roughness was the lowest for specimens with higher SiC content and decreased with the increase in direction angle. The mechanism of interaction with whiskers at different angles is presented in Figure 6.29.

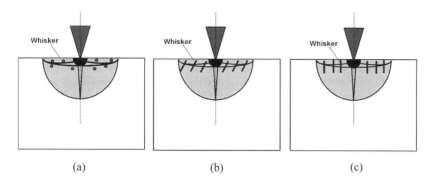

Figure 6.29. *Schematic illustration of the interactions between the abrasive particles and the whiskers of specimens when (a) θ = 0; (b) θ = 45 and (c) θ = 90 surface are used as machining surface [LEE 01]*

Depending on the angle of the whisker plane to crack propagation, whiskers have different effects on the material characteristics. For the 0° angles there is no toughening effect and the crack easily propagates through the matrix. For 45° angles, whiskers experience tension and bending and, due to the bending effect, the fracture of whiskers occurs. For the 90° angles, whiskers were able to withstand higher stresses than any other cases.

6.6. Application of CMCs: cutting tool inserts

Chemical processes used to actually synthesize the powders have crucial roles in the characteristics of the final material [MON 05]. The synthesis used in this paper is the synthesis of reactive powders of advanced composites of alumina reinforced with yttria partially stabilized zirconia (Y-PSZ) by EDS processing using an alcoholic solution of $ZrOCl_{2.8}H_2O$ and $Y(NO_3)_3$. The inserts created were tested using a Vickers hardness test for hardness and machining tests have been conducted to determine flank wear, crater wear, tool life and chipping resistance at different machining parameters. Owing to the fact that flank wear greatly influences dimensional stability in turning processes, the influence of cutting speed on flank wear has been investigated and presented in Figure 6.30. It can be seen that there is an increase in flank wear with cutting speed.

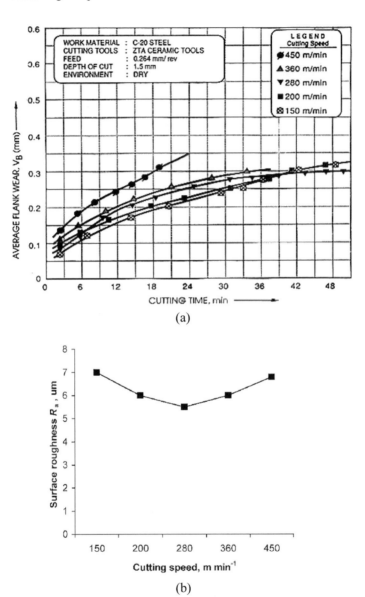

Figure 6.30. *Variation of (a) average flank wear with machining time at different cutting speeds and (b) surface roughness (Ra) with cutting speed [MON 05]*

Surface quality of the workpieces turned with CMC inserts has been analyzed through analysis of the cutting nose of the tool insert. It is shown in Figure 6.30b that the cutting speed greatly influences surface roughness and shows that there is an optimum value of the speed where the surface roughness has a minimum value. This can be explained by the increase in tool wear after 250 m min^{-1}. It has been stated that stress and high temperature during high-speed machining induce phase changes. Fracture and chipping occur at the corners as well as crater surface due to exceeding the tool material's tensile strength. The wear is the consequence of the plastic deformation of alumina and the adhesive wear mechanism where steel chips on the rake face adhere to the material and pull out alumina and zirconia grains. Plastic deformation occurs due to the chip flow [MON 05]. Cutting experiments have been conducted to determine the effect of adding yttrium to reinforce a Al_2O_3/Ti(C,N) ceramic matrix composite. Flexural strength, fracture toughness, fracture resistance and wear resistance have been determined and discussed [XU 01]. The effect of yttrium on mechanical properties of the CMC is presented in Figure 6.31. It can be seen that there is an optimal amount of yttrium that yields maximum value of flexural strength and fracture toughness. It can be seen that the optimal value is at 0.5% yttrium when the phases are homogenously distributed in the ceramic matrix with grain size about 1 to 2 µm. The coarsening of the particles at higher percentages of yttrium causes a decrease in the mechanical properties of the material. The hardness of the material did not seem to have been significantly affected by the change in amount of reinforcing phase.

Figure 6.32 shows the wear resistance of CMC materials with different amouts of Y, TCN1 – 0.5% Y, TCN2 – 2% Y, TCN0 – 0% Y. It can be seen that the material with 0.5% Y has the highest flank wear, while the material with 2% Y has the lowest tool wear. This is in agreement with the previously observed mechanical properties. When analyzing the failure mechanisms, it has been determined that there is crater wear, flank wear, and the slight edge chipping. Adhesion wear and ploughing have been detected in the flank wear. The flank wear mechanism changes from abrasive wear to adhesive wear with the increase in cutting speed, while the crater wear increases due to the fact that the adhesion between the tool and the material is increased. Tool fracture is observed in the case of intermittent cutting.

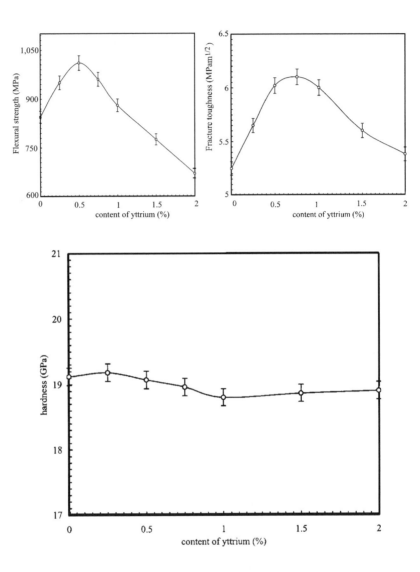

Figure 6.31. *Effects of yttrium on the mechanical properties of Al$_2$O$_3$/Ti(C,N) ceramic tool materials [XU 01]*

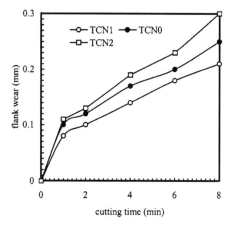

Figure 6.32. *Flank wear curves of $Al_2O_3/Ti(C,N)$ series ceramic tool materials when turning hardened #45 steel. TCN1 – 0.5%Y, TCN2 – 2%Y, TCN0 – 0%Y [XU 01]*

Figure 6.33 shows the difference in fracture resistance for the three materials investigated. It has been suggested that there is a strong influence of cutting parameters on the form of fracture. In case A, the predominant mechanism of failure is peeling in the rake face area and tool fracture in the flank area while for the two other sets of materials, the main mechanism is tool fracture in the flank area and thermal fracture in the rake face area.

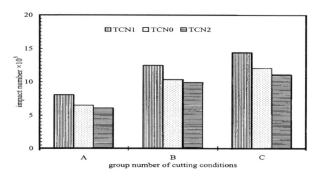

Figure 6.33. *Fracture resistance of $Al_2O_3/Ti(C,N)$ series ceramic tool materials, group A: cutting speed v = 118 m/min, feed rate f = 0.1 mm/rev, depth of cut ap = 0.5 mm; group B: v = 188 m/min, f = 0.1mm/rev, ap = 0.3 mm; and group C: v = 264 m/min, f = 0.1 mm/rev, ap = 0.1 mm [XU 01]*

This implies that the optimal wear and fracture resistance can be achieved by the increase of mechanical properties through addition of the optimal amount of Y reinforcing phase [XU 01]. Szafran et al. [SZA 00] investigated the characteristics and tool life of the Si_3N_4-Al_2O_3-TiC-Y_2O_3 system used for cutting blades. Three materials have been investigated in the study, A- 60% Si_3N_4, 10% Al_2O_3 + Y_2O_3, 30% TiC, B- 60% Si_3N_4, 20% Al_2O_3 + Y_2O_3, 20% TiC, C- 60% Si_3N_4, 30% Al_2O_3 + Y_2O_3, 10% TiC. The TiC phase has been added to act as the reinforcing element in the ceramic composite, increasing hardness and the stress intensity factor, K_{IC}.

Wear of the cutting blades was examined during machining operations and determined by measurement of wear depth on the flank face. For comparison, the material has been compared to commercially available Si_3N_4-Al_2O_3-MgO (Figure 6.34).

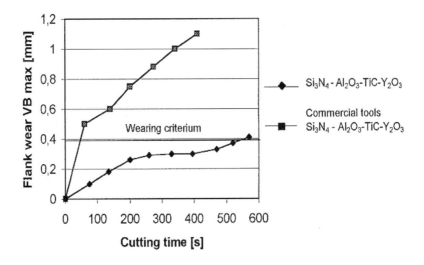

Figure 6.34. *Flank wear vs. cutting time [SZA 00]*

It can be seen that the flank wear of the plate C is significantly lower compared to Si_3N_4-Al_2O_3-MgO throughout the entire cutting process. Temperature in the cutting area has been measured using a pyrometer and presented as a function of time (Figure 6.35).

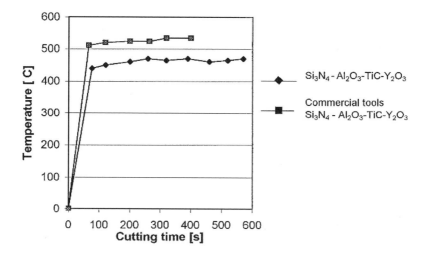

Figure 6.35. *Temperature within the machining region vs. cutting time [SZA 00]*

It is obvious from the results presented that the temperature within the machining region is lower for the material investigated when compared to the commercially available Si_3N_4-Al_2O_3-MgO.

6.7. Review of various technologies for machining CMCs

Tuersley *et al.* [TUE 94] reviews a range of technologies used for machining CMCs such as grinding, ultrasonic machining, abrasive water jet machining, EDM and laser machining.

Grinding is a necessary operation for finishing components and although it is necessary it introduces defects into the material which are extremely difficult to detect due to their small size. One of the grinding methods for these materials is ductile regime grinding where, by minimizing grinding forces through constant material removal rate, we minimize forces and thus the generation of cracks. Another issue of grinding materials like ceramics is the ratio of tangential and normal forces which induce elastic deformation of machine parts like spindles and wheels. In these cases it is considered necessary to use "stiff" machines and extremely precise set-ups to compensate for these issues. Owing to the time-consuming characteristic of ductile regime grinding, creep-feed grinding has been considered to be one of the options for machining materials like CMCs. This method differs from the ductile grinding by moving the table at very low speeds while using the full grind depth. Naturally, there is no ideal set-up and every CMC will need analysis to determine the ideal parameters.

Ultrasonic machining involves a combination of vibration and abrasives which results in micro-grinding of the material. Usually for ceramic materials, boron carbide in water suspension has been chosen as the abrasive which yields the best results. The grit size of the particles has a detrimental effect on the surface finish. Changing the grit from 600 to 240 can yield the change in finish from 12 to 25 microns. Ultrasonic machining has also been combined with other conventional techniques, such as rotary machining where the slurry does not cut but provides a cleaning action which eliminates build-up, binding, and loading of the diamond matrix providing higher efficiency in the process. Ultrasonic machining has been combined with non-conventional machining techniques such as electro-discharge grinding where the grinding forces are lowered and removal rates increased.

Abrasive water jet also uses abrasive particles and is mostly used for slots and grooves, but lately has been reported as being used for turning, milling and drilling. The most influential factors are considered to be pressure, distance of the nozzle from the work piece, abrasive type, flow rate, material, geometry of the nozzle, etc. When machining CMCs the pressure needs to be adjusted to avoid fracture, cracking and delamination due to the brittle nature of the material itself. Nozzle materials are an issue due to the fact that highly abrasive materials are used and can wear out a nozzle within minutes. Flow rates usually have a critical value for the efficiency, yielding a decrease if the critical value is surpassed. Surface quality has been addressed as well showing that the process is not completely free of thermal effects and it can induce changes in the area 500 μm around the hole entry as well as a rougher surface and limited plastic deformation when using higher speeds. In CMCs, the material removal rates are up to 20 times greater than those for diamond saw cutting. However, the surface roughness produced is up to ten times higher.

EDM is a technology that can only be applied to conductive materials. EDM uses either evaporation or thermal spalling as techniques for material removal. Factors like polarity, pulse duration and pulse current have a great effect on the final properties of the machined surface. The advantages of the process include the ability to produce complex shapes, the absence of mechanical stresses, high MRR, etc. The biggest disadvantage of the process is the limitation to conductive materials. The size and type of the added phases is critical since different effects regarding crack deflection and strengthening, will be detected.

Laser machining has the advantage that there is no tool to be used and therefore no tool deflection or vibrations can affect the process creating the possibility of making parts of extremely complex design. There are no abrasives and therefore damage to the surface is avoided. The crucial parameters considered in a laser machining process are thermal characteristics of the workpiece material, manner of heat application (pulsed versus continuous wave), efficiency of removing metal from the cutting path, etc. Owing to the fact that ceramic materials have low thermal conductivity and therefore easily develop micro-cracks, spalling and even failure, the power density, peak power and pulse characteristics have a critical influence on their machinability. Formation of the re-cast layer on the surface as well as the formation and avoidance of micro-cracks is one of the topics often explored in the area of CMCs. Re-cast layers have been eliminated using YAG lasers when submerged under water although this is only applicable to ceramics that undergo sublimation. For ceramics with a very high melting point, the re-cast layer cannot be avoided. Using lasers on reinforced matrices results in larger heat-affected zones but also less damage than when machined with diamond saws.

6.8. References

[CAR 00] CARROLL, J.W., TODD, J.A., "Laser machining of ceramic matrix composites", in *Ceramics Engineering and Science Proceedings*, 2000, The American Ceramics Society, P. 456 - 466.

[FU 94] FU, J.M., LIU, D.M., "The effect of electrodischarge machining on the fracture strength and surface microstructure of an Al_2O_3/Cr_3C_2 composite", *Material Science and Engineering*, 1994, 188, p. 91-96.

[GON 94] GONCZY, S.T., LARA-CURZIO, E., REISTER, L., BUTLER E.P., DANFORTH, S.C., CANNON, W.R., "Low cost machining of ceramic-fiber reinforced ceramic composites (2-D) and the effects on the tensile properties", *Processing, Design and Performance of Composite Materials*, ASME, 1994, 52, p. 217-238.

[HOC 00] HOCHENG, H., TAI, N.H., LIU, C.S., "Assessment of ultrasonic drilling of C/SiC composite material", *Composites: Part A*, 2000, 31, p. 133-142.

[JIA 02] JIANXIN, D., TAICHIU, L., "Ultrasonic machining of alumina-based ceramic composites", *Journal of the European Ceramic Society*, 2002, 22, p. 1235-1241.

[JON 01] JONES, A.H., TRUEMAN, C., DOBEDOE, R.S., HUDDLESTON, J., LEWIS, M.H., "Production and EDM of Si3N4-TiB2 ceramic composites", *British Ceramic Transactions*, 2001, 100(2), p. 49-54.

[LAU 08] LAUWERS, B., BRANS, K., LIU, W., VLEUGELS, J., SALEHI, S., VANMEENSEL, K., "Influence of the type and grain size of the electro-conductive phase on the Wire-EDM performance of ZrO_2 ceramic composites", *CIRP Annals - Manufacturing Technology*, 2008, 57, p. 191-194.

[LEE 01] LEE, T.C., JIANXIN, D., "Ultrasonic erosion of whisker-reinforced ceramic composites", *Ceramics International*, 2001, 27, p. 755-760.

[LI 05] LI, Z.C., JIAO, Y., DEINES, T.W. PEI, Z.J., "Rotary ultrasonic machining of ceramic matrix composites: feasibility study and designed experiments", *International Journal of Machine Tools & Manufacture*, 2005, 45, p. 1402-1411.

[MON 05] MONDAL, B., "Zirconia toughened alumina for wear resistant engineering and machinability of steel application", *Advances in Applied Ceramics*, 2005, 104 (5), p. 256-260.

[PIT 00] PITMAN, A., HUDDLESTON, J., "Electrical discharge machining of ZrO_2/TiN particulate composite", *British Ceramic Transactions*, 2000, 99 (2), p. 77-84.

[PUT 01] PUT, S., VLUEGELS, J., VAN DER BIEST, O., TRUEMAN, C.S., HUDDLESTON, J., "Die sink electrodischarge machining of zirconia based composites", *British Ceramic Transactions*, 2001, 100 (5), p. 207-213.

[SZA 00] SZAFRAN, M., BOBRYK, E., KUKLA, D., OLSZYNA, A., "Si_3N_4-Al_2O_3-TiC-Y_2O_3 composites intended for the edges of cutting tools", *Ceramics International*, 2000, 26, p. 579-582.

[TUE 94] TUERSLEY, I.P., JAWAID, A., PASHBY, I.R., "Review: Various methods of machining advanced ceramic materials", *Journal of Materials Processing Technology*, 1994, 42, p. 377-390.

[TUE 98] TUERSLEY, I.P., HOULT, T.P., PASHBY., I.R., "The processing of SiC/SiC ceramic matrix composites using a pulsed Nd-YAG laser, Part I: Optimisation of pulse parameters", *Journal of Materials Science*, 1998, 33, p. 955-961.

[WU 94] WU, J.M., LIU, D.M., "The effect of electrodischarge machining on the fracture strength and surface microstructure of an Al_2O_3/Cr_3C_2 composite", *Material Science and Engineering*, 1994, 188, p. 91-96.

[XU 01] XU, C., HUANG, C., AI, X., "Mechanical property and cutting performance of yttrium-reinforced Al2O3/Ti(C,N) composite ceramic tool material", *Journal of Materials Engineering and Performance*, 2001, 10(1), p. 102-107.

List of Authors

Alexandre M. ABRÃO
Department of Mechanical Engineering
Federal University of Minas Gerais-UFMG
Belo Horizonte Minas Gerais
Brazil

Juan C. CAMPOS RUBIO
Department of Mechanical Engineering
Federal University of Minas Gerais-UFMG
Belo Horizonte Minas Gerais
Brazil

François CÉNAC
JEDO Technologies
Labège
France

Francis COLLOMBET
University of Toulouse
 INSA, UPS, Mines Albi, ISAE, ICA
(Institut Clément Ader)
Toulouse
France

J. Paulo DAVIM
Department of Mechanical Engineering
University of Aveiro
Portugal

Michel DÉLÉRIS
JEDO Technologies
Labège
France

Paulo E. FARIA
Department of Mechanical Engineering
Federal University of Minas Gerais-UFMG
Belo Horizonte Minas Gerais
Brazil

Frank GIROT
Faculty of Engineering of Bilbao
University of the Basque Country
Bilbao
Spain

Mª Ester GUTIÉRREZ
Faculty of Engineering of Bilbao
University of the Basque Country
Bilbao
Spain

Daniel ILIESCU
Arts et Métiers ParisTech, LAMEFIP
Talence
France

Mark J. JACKSON,
Purdue University
Center for Advanced Manufacturing, College of Technology
West Lafayette Indiana
USA

Aitzol LAMIKIZ
Faculty of Engineering of Bilbao
University of the Basque Country
Bilbao
Spain

Luis Norberto LÓPEZ DE LACALLE
Faculty of Engineering of Bilbao
University of the Basque Country
Bilbao
Spain

Tamara NOVAKOV
Purdue University
Center for Advanced Manufacturing, College of Technology
West Lafayette Indiana
USA

Alokesh PRAMANIK
School of Mechanical and Manufacturing Engineering
University of New South Wales
Sydney
Australia

Liangchi ZHANG
School of Mechanical and Manufacturing Engineering
University of New South Wales
Sydney
Australia

Rédouane ZITOUNE
University of Toulouse
INSA, UPS, Mines Albi, ISAE, ICA
(Institut Clément Ader)
Toulouse
France

Index

A, C

abrasive, 167
  machining, 35, 179, 180
  water jet machining, 194
abrasive water jet (AWJ), 167-178,
ceramic matrix composites (CMC), 213, 234, 241, 242, 247, 252
chemical damage, 66
curing conditions
  effect, 12
cutting, 167-178
  forces, 7
  force modeling, 2, 7, 8, 10, 13, 16, 19, 20, 36
  mechanisms, 54, 58, 60, 64, 105
  orthogonal, 1, 2, 3, 14, 21
  parameters, 123, 133, 145, 151-157
  tool inserts, 213, 245

D, E

delamination, 115, 119, 120, 122, 124, 131, 132, 136, 140- 153, 158, 173
dimensional and geometric
  deviations, 113, 153, 159
drilling, 1, 2, 21-35, 182, 184, 185, 186, 191
  forces, 113, 157-159
  technology, 113
electro-chemical machining, 190, 193, 194, 203
electro-discharge machining, 203, 213

F–K

FE simulation, 202
fiber
  nature, 64
  orientation, 54-63, 66
GFRP routing, 70
grinding, 182, 187-189
jet, 167
kerf
  quality, 172
KFRP routing, 89

L, M

laser machining, 213, 227, 233, 234, 237, 251, 253
laser-beam machining, 192

long fibers, 21
machinability, 39, 106, 107
machining
    conventional, 182
    non-conventional, 190, 203
mechanical damage, 65, 66, 71, 90
metal matrix composites (MMCs), 181-195, 202, 203
milling, 168, 178-182, 189
    of CFRP, 75
    of composite materials, 70

P, S

plastic zone, 199
polymer matrix composites reinforced by long fibers (PMCRLF), 1, 2, 3, 9, 21, 22, 25, 28-36
polymer matrix composites, 21
special tools, 117
stress field, 196, 197
subsurface damage, 7, 8, 11, 35
surface
    integrity, 140, 159
    quality, 72, 74, 105
    roughness, 4-12, 36, 122, 149-153, 158

T

technologies for machining CMCs, 251
thermal damage, 65, 69, 79, 90
tool, 41, 76, 84, 86, 87, 90, 110
    geometries, 70, 88, 89, 101
    materials, 41, 42
    types, 88
    wear, 39, 43, 47, 49, 59, 73, 74, 78-99, 117, 119, 124-139, 144, 148, 153, 154, 155, 157, 158
tool–workpiece interaction, 195
turning, 182, 189, 203
    conditions, 103, 105
    of composite materials, 101

U, W

ultrasonic machining, 213, 234, 237-241, 251, 252
water, 167
water jet machining (WJM), 226, 227, 237, 251
wear mechanism, 171
wedge mechanism, 170